CONSERVING DATA IN THE CONSERVATION RESERVE

CONSERVING DATA IN THE CONSERVATION RESERVE

How a Regulatory Program Runs on Imperfect Information

JAMES T. HAMILTON

Washington, DC • London

First published in 2010 by RFF Press, an imprint of Earthscan

Earthscan LLC,1616 P Street, NW, Washington, DC 20036, USA
Earthscan Ltd, Dunstan House, 14a St Cross Street, London EC1N 8XA, UK
Earthscan publishes in association with the International Institute for Environment and Development

For more information on RFF Press and Earthscan publications, see www.rffpress.org and www.earthscan.co.uk or write to earthinfo@earthscan.co.uk

ISBN: 978-1-933115-82-5 (hardback)
ISBN: 978-1-933115-81-8 (paperback)

Copyedited by Kristin Hunter
Typeset by Andrea Reider
Cover design by Ellen A. Davey
Cover image from U.S. Dept. of Agriculture

Library of Congress Cataloging-in-Publication Data

Hamilton, James, 1961-
Conserving data in the Conservation Reserve : how a regulatory program runs on imperfect information / James T. Hamilton.
p. cm.
Includes bibliographical references and index.
ISBN 978-1-933115-82-5 (alk. paper) — ISBN 978-1-933115-81-8 (alk. paper)
1. Conservation Reserve Program (U.S.) 2. Agriculture and state--United States.
I. Title.
HD1761.H356 2010
354.3'3427940973—dc22 2009047025

A catalogue record for this book is available from the British Library

At Earthscan we strive to minimize our environmental impacts and carbon footprint through reducing waste, recycling and offsetting our CO_2 emissions, including those created through publication of this book. For more details of our environmental policy, see www.earthscan.co.uk.

Printed and bound in the UK by TJ International.
The paper used is FSC certified.

About Resources for the Future *and* RFF Press

Resources for the Future (RFF) improves environmental and natural resource policymaking worldwide through independent social science research of the highest caliber. Founded in 1952, RFF pioneered the application of economics as a tool for developing more effective policy about the use and conservation of natural resources. Its scholars continue to employ social science methods to analyze critical issues concerning pollution control, energy policy, land and water use, hazardous waste, climate change, biodiversity, and the environmental challenges of developing countries.

RFF Press supports the mission of RFF by publishing book-length works that present a broad range of approaches to the study of natural resources and the environment. Its authors and editors include RFF staff, researchers from the larger academic and policy communities, and journalists. Audiences for publications by RFF Press include all of the participants in the policymaking process—scholars, the media, advocacy groups, NGOs, professionals in business and government, and the public.

CONTENTS

ACKNOWLEDGMENTS

Once information is created, the marginal cost of circulating it is almost zero in the age of the internet. A central quandary of the information age, however, is who is willing to pay for the initial creation of an idea, a study, or a story. This book owes its existence to the generous support of a fellowship in environmental regulatory implementation provided by Resources for the Future (RFF). This program was in turn funded by a grant to RFF by the Andrew W. Mellon Foundation, which aimed to increase understanding of the regulatory process by funding studies of how particular regulations came to be. Additional support came from the Center for Advanced Study in the Behavioral Sciences, which provided me with an unparalleled research environment during a year's sabbatical from Duke University.

I owe a great debt to Ben Hendricks, who worked as a research assistant on this book during his years as a Duke undergraduate. Ben conducted interviews, tracked down government documents, and even filed Freedom of Information Act requests to find out about implementation decisions made in the Conservation Reserve Program (CRP). He is a naturally gifted researcher, and I look forward to seeing his own work published some day. Kim Krzywy did wonderful work on this project for three years, starting with the grant proposal that resulted in RFF support and ending with the proofing of the manuscript. Her attention to detail and willingness to track down hard-to-find information has made many chapters more informative.

I also am indebted to many people for the insights they provided on the operation of the CRP, including Jessica Munn, Janine Pierson, Dan Hellerstein, Alex Barbarika, Robert Stephenson, Mike Linsenbigler, Parks Shackelford, Paul Harte, Ralph Heimlich, Tim Osborn, J. Timothy Jones, Susan O. Woodall, and Miles B. Davis. I especially appreciated the willingness of government officials involved in the implementation of the CRP in Washington, DC, and in field offices to answer questions about the program's evolution through telephone and in-person interviews. Further, Don Reisman of RFF and a set of outside reviewers provided highly valued feedback on the manuscript.

This book is dedicated to my grandfathers. James Towler began his working life as a county agricultural agent and eventually moved to a small town to open up a hotel. Henry Hamilton started out as a one-room school teacher who loved to cultivate his own crops and eventually moved "down the valley" to become a high school principal. The gifts and love they gave their children and grandchildren continue to be felt today.

James T. Hamilton

FOREWORD

I remember feeling rather stunned as I looked out early one morning across a broad expanse of tall prairie grass billowing in the breeze over the rolling hills of Chouteau County, in north central Montana. As best I can recall, it was sometime in1989 or 1990. And someone in the small contingent of conservationists touring the county that day matter-of-factly remarked to the group: "That's CRP."

That's the Conservation Reserve Program? I marveled. *Well I'll be damned. It worked!*

I kept that thought to myself that morning. It wouldn't have done my reputation as a lobbyist much good to exclaim surprise that a multi-billion dollar program I'd been pushing for the better part of a decade actually seemed to work once it was implemented.

After all, I'd had a hand dreaming up the CRP in the early 1980s. I'd crunched piles of numbers to make the case that long-term land retirement was the only practical way to significantly curb the serious loss of topsoil from the nation's most fragile, privately owned cropland. I'd written extensively and testified repeatedly before congressional committees, confidently extolling the promise of such a program and urging its inclusion in the 1985 farm bill. I'd walked the halls of Congress, the Office of Management and the Budget, and the Department of Agriculture, armed with fact sheets and analyses and earnest memos, and haggled over regulatory details as the program got underway.

But I did indeed find myself taken aback that morning. We weren't just looking at a field of beautiful native grasses. We were looking at a *government program*. A federal law and the rules implementing it had *caused* that grass to be sown where wheat had grown for decades before. Erosion-depleted soils were being rebuilt and replenished under that protective blanket, which was also a boon to wildlife.

At that point the dramatic expansion of the CRP was *the story* in farm country, and the rare good story, at that. And it had been invented and lobbied into existence by a bunch of environmentalists who refused to believe they couldn't make the vision of a vast land reserve a reality.

As my friend, the great conservationist Max Schnepf (then and for many years an executive with the Soil and Water Conservation Society) and I drove around Chouteau county that day, Conservation Reserve Program ground was everywhere, the grass in various stages of maturity, alternating with wheat fields farmers had not (yet!) converted to the program, and pasture and range that didn't qualify. The "Food Security Act" that had created the CRP had only been on the books a few years—since December, 1985—and the first "sign up" for the program the following year had made all of us who had worked to invent the reserve extremely nervous. Only about 2 million acres nationwide came into the CRP, and the statute called for up to 45 million.

But the second year saw more than 13 million acres added to the reserve, and by 1988 nearly 24 million acres were enrolled. In Chouteau, CRP acreage exploded from under 1,000 to over 107,000 in that brief interval—and Chouteau was by no means the most heavily enrolled county in the land. Eventually CRP spending would reach $2 billion per year, and at its peak encompass over 35 million acres (32 million acres are contemplated in the most recent, 2008, farm bill, the ethanol boom notwithstanding).

In *Conserving Data in the Conservation Reserve*, James Hamilton has meticulously and thoughtfully described how the CRP was implemented, and continually improved, as analysts and program managers and land owners and Congress assembled and sorted through constantly changing and improving information that nonetheless always was, and always will be, imperfect. The book focuses on information and analysis that led to improvements in the program after 1990, with the introduction of an Environmental Benefits Index to focus program enrollment on land that would yield greater environmental benefits per dollar expended.

Fittingly enough, a great big pile of imperfect information, unprecedented in its nature and scope, and made accessible to analysts by new information technology, is exactly how the CRP got its start in the late 1970s. In fact, the CRP has always been about data, its generation, use, and, in Hamilton's sense, its conservation.

***

By the end of the Carter administration, two unprecedented streams of information and analysis were pointing to severe shortcomings of traditional soil conservation policies and programs.

First were the data and ideas that emerged from a series of internal USDA evaluations of the department's two primary conservation programs, both of them voluntary and decades old. The Agricultural Conservation Program had provided farmers financial "cost-sharing" assistance and was administered by USDA's subsidy arm (the Agricultural Stabilization and Conservation Service, whose functions are now subsumed under the Farm Services Agency). The Conservation Technical Assistance Program offered farmers free conservation planning and implementation services, and was administered by thousands of field personnel of the Soil Conservation Service (SCS, now known as the Natural Resource Conservation Service).

Second were the compelling insights that a small group of analysts extracted from a database, massive in its day, that had been generated in 1977 by an unprecedented on-the-ground survey of America's private farmland—the first National Resources Inventory, conducted by the Soil Conservation Service.

The information that emerged from those sources merely attached numbers to what anyone with an eye for conservation could readily see in those days on America's agricultural landscape. But it was the numbers that brought about the historic changes in conservation policy of the early-to-mid 1980s.

When farmers responded to Agriculture Secretary Earl Butz's admonition in the early 1970s to "feed the world" by planting corn, wheat and soybeans "fencerow to fencerow" across the countryside in response to what turned out to be a short-lived run-up in food prices, the impact on the land was devastating. Butz's policy prescriptions had a disastrously brief shelf-life that gave way, by 1975, to plunging commodity prices and farm incomes, a surge in suicides across farm country, and angry "tractorcade" protests of farmers plowing up the Washington Mall. The Butz boom-to-bust program also triggered the destruction of an entire generation of soil conservation practices within the space of a few years, as farmers cleared and plowed every acre in sight. In 1979, the journalist James Risser won a Pulitzer Prize (his second) for an incredible seven-part series in the *Des Moines Register* documenting this environmental disaster down on the farm, and soil erosion was a featured part of his investigation.

The ACP program, the evaluation showed, was in balance an embarrassing, multi-billion-dollar conservation failure, and likely had been for decades. Cost-share funds were distributed across the country and within states via formulas that had everything to do with politics and nothing to do with conservation needs on the land. Farmers were doled out cost-share payments on a first-come, first-served basis at local USDA offices with no evaluation criteria and no prioritization to direct the program to farmers and farm fields with the most serious soil conservation problems. Some of the grants amounted to a hundred bucks, the conservation equivalent of election-year walk-around money that curried favor but probably accomplished very little conservation.

To put the evaluation on a solid, not to say innovative, technical foundation, the man who designed it, a great conservationist with ASCS, Gordon Nebeker, had made sure field staff incorporated a new technique for estimating soil loss—the Universal Soil Loss Equation—so that they were able to predict soil erosion rates before and after the ACP practices were implemented. If the program had been targeted to the fields that most needed it and the most cost-effective practices, Nebekers's evaluation showed, the amount and the cost-per-ton of erosion prevention would dramatically improve. But without a major targeting effort, a tremendous amount of taxpayer money would be wasted in the name of conservation.

The Conservation Technical Assistance (CTA) evaluation was devised by another great conservationist, the SCS economist Roy "Mack" Gray. Like Nebeker over at ASCS, Gray strongly believed that their programs weren't getting enough bang for the taxpayers' hundreds of millions of bucks and designed his evaluation work to make things better. The CTA evaluation basically showed that under the management and reward systems then in place, most SCS personnel saw their job as preparing elaborate, "whole farm" conservation plans, not getting specific, cost-effective conservation practices on the ground. For the most part the SCS had no money to offer farmers—that was under the control of the ACP—and precisely because they were so complex, most of those conservation plans sat on shelves, one copy at the SCS office, the other at the farm, gathering dust. And some of that dust was coming off tilled fields that remained unprotected from erosion despite decades of financial and technical assistance to farmers.

So the two main programs USDA deployed to combat soil erosion, protect water quality and improve wildlife habitat were no match for the environmental calamities Jim Risser and so many others were documenting on the land. Some new tools were badly needed.

The first time I heard the *idea* that would become the CRP was in 1981 or 1982, from the agricultural economist Charles Benbrook, at the time of writing the chief scientist for The Organic Center, the small but influential think tank of the organic food and agriculture industry. For my money, Benbrook would have to be considered the single most influential policy analyst of my generation working at the nexis of agriculture and environment, on topics as wide-ranging as commodity program reform, pesticide policy, genetically modified crops, organic agriculture, and the use of antibiotics in farming. But some of his most important contributions to agri-environmental policy came from his work in the early 1980s with soil erosion and conservation data pried out of the 1977 NRI.

The power of the NRI database derived from the fact that SCS staff had gone to tens of thousands of statistically-determined sampling points in fields across the country and at each spot had recorded detailed information about the resource base. In particular, SCS field experts had recorded values for each of the individual factors in the USLE—even more detail than Gordon Nebeker's evaluators had documented for land in the ACP. (C was the crop cover value; R was rainfall; K was soil erodibility; L and S were length and steepness of slope, respectively.) Erosion had never been estimated by the SCS in this way before, and agency personnel also recorded on their elaborate forms the presence or absence of any conservation measures. Those paper records were then transposed into one large, "machine readable" database. Readable, that is, if by chance you had access to one of that era's room-sized, "mainframe" computers.

Miller, Benbrook, and the renowned soil scientist William Larson at the University of Minnesota were able to ask some very important new questions about patterns of soil erosion in the United States using the unprecedented information captured in the NRI database. I pitched in too, as reams of NRI printouts, many of them bootlegged by Miller, who had access to the USDA mainframe computer on which NRI data were stored, found their way to me, forming the core information I used in 1983 and 1984 to write the American Farmland Trust report, "Soil Conservation in America: What Do We Have To Lose?" (AFT, 1985). I laboriously retyped data from hundreds of tabulations into an ingenious device called a "desktop computer" equipped with a powerful "software program": imagine a spreadsheet, only it's *electronic*! To flip on the attached, thundering "daisy-wheel printer" that churned out drafts of the AFT report was to experience, in some small measure, the thrill of being strafed by a fighter plane in World War II.

The NRI analyses revealed that soil conservation was highly concentrated: the majority of the soil "loss" or movement recorded in the NRI was associated with about 10 percent of the most erodible cropland that had extremely high erosion rates. Controlling erosion on the most fragile soils could thus dramatically reduce total soil displacement, and cost-effective conservation farming practices (like the newly emerging reduced tillage systems) could handle much of the rest. Further analysis found that conservation practices were rarely in place on the most erosion-prone land despite decades of taxpayers' financial aid provided through the ACP, and just as many decades of free technical assistance provided by thousands of SCS field staff. If important changes weren't made in how conservation policy was conceived and implemented, there was no chance that the programs in place would measurably reduce the country's soil erosion problems.

Butz's ill-considered effort to "get the government out of agriculture" in the early 1970s quickly gave way to a prolonged period of more government involvement in farming than ever before. After a few years of tight supplies, the country faced annual, price-depressing crop surpluses that cost taxpayers and farmers billions. USDA combated those mountains of excess commodities with equally massive, annual "set-aside" requirements that idled as much as 70 million acres of cropland in some years, as farmers were required temporarily to remove from production 10 percent, 15 percent, or more of the acreage they had enrolled in the commodity programs as a quid pro quo for getting subsidy payments on their remaining production. It didn't matter if they were farming in the rolling, erosion-scarred hills of northern Missouri or the pancake flat, highly productive soils of central Illinois: everyone had to idle land if they wanted to qualify for crop subsidies, without which few farmers could continue to operate.

Benbrook and Miller saw in this conundrum of excessive production and equally excessive policies to control it, the makings of a grand conservation solution informed by the NRI. Why not aim the land retirement features of the commodity programs at the most erodible acres, and replace the year-to-year set-aside program that brought few conservation benefits (and often harmed both soils and wildlife) with a program of longer term (mostly 10-year) contracts, during which the land would be protected by grass or trees? The old Soil Bank of the 1950s was an obvious model for the program. But its 10-year land retirement contracts for millions of acres of grass and treeplantings (long since brought back into production) had not been directed specifically to fragile land. The NRI analyses demonstrated the policy opportunity and the parameters for targeting, and the USLE (and a similar equation for wind erosion) provided a tool to make the policy operative on the ground. All we needed were billions of dollars a year to plant the grass and trees where, in our best judgment at the time, it would do the most good.

The driving force behind the AFT report was the man who hired me to write it and who gave the CRP its legs as a policy proposal: Bob Gray. Part bulldog and part impresario, Gray was a vice president at AFT who tirelessly arm twisted and cajoled Congress to make the program a reality. There is no question in my mind that the CRP would not exist today if it hadn't been for the skill and tenacity of Bob Gray in the mid-1980s, including his leadership of the "conservation coalition" that met weekly throughout the 1985 farm bill debate to plot strategy and direct lobbying efforts.

It's been said that the 2008 farm bill attracted unprecedented attention to farm policy from reformers outside mainstream agriculture, but it's been said mostly by people for whom 2008 was their first farm bill. The 1985 farm bill attracted by far the most potent array of "outsiders" I've seen in the seven farm bills I've worked on going back to 1977. The large coalition Gray assembled included a major involvement from the National Audubon Society, in the person of the formidable Maureen Hinkle (who taught me how to lobby) and the tireless Dan Weiss of the Sierra Club. Both of them, and many others, mastered all the information we analysts were churning out, and added the true grassroots clout and shoe-leather lobbying that got the CRP and other historic conservation provisions included in the 1985 farm bill.

Most of us testified about these ideas more than once before Congress during 1984 and 1985, and we were very fortunate to have as our prime audience two legislators who understood the analytical case we were making and embraced it. Chairman Ed Jones, Democrat of Tennessee, and his two aides, Bob Cashdollar and Jimmy Johnson, championed the landmark conservation title of the farm bill in the House. Senator Richard Lugar, the Indiana Republican, and his subcommittee staff director, Chuck Conner (Undersecretary of Agriculture during the 2008 farm bill) were our stalwarts in the Senate.

****

I thought of that morning in Montana twenty-some years ago as I read James Hamilton's closing words about another parcel of CRP ground, this one in North Carolina, which inspired him to tell the bureaucratic history of the program at the beginning of the book through the experience of its owner, "James Grass":

> Even without full information about what happened, if legislators who voted for the CRP and FSA officials who helped develop the Environmental Benefits Index showed up at this North Carolina farm, they would appreciate and approve of the soil and wildlife conservation benefits there that are due to the Conservation Reserve Program.

That's certainly been my experience over the years. Hamilton's insightful excursion through the generation and use of information in the implementation of the nation's largest conservation program has at its core important insights about the way policymakers try to balance the technocratic demand for ever better, ever more granular information about the design, performance and impact of government programs, against the costs—monetary and political—of generating that information and making it available to everyone with a dog in the fight. The title of the book derives from Hamilton's observation that throughout the development and operation of a major federal program like the CRP, those involved find themselves wanting and demanding more program information, even as they act to conserve the amount and quality of facts and analyses needed to meet the objectives Congress intended when creating it. Bureaucracies concoct program rules and regulations, and with a modicum of evaluation and oversight they are entrusted to make crucial decisions in the absence of anything approximating perfect information about program performance.

To do otherwise, to make imperfect information the enemy of the good, Mr. Hamilton observes, runs the risk of dooming programs to excessive administrative costs, stress on employees who must both deliver program performance and report on it, and attrition of political support. The first generation of information underlying the CRP gave way to the Environmental Benefits Index, which itself has evolved over time as new information begat more ambitious performance goals, and new performance goals beget new information . . . and so on.

The real question about the CRP is not how imperfect it is. Surely taxpayers have spent too much money protecting land that should never have been enrolled in the CRP in the first place, and ought to be back in crop production now. Just as surely, we could also do a far better job enrolling land that contributes the most to water pollution—which is often not the land with the highest erosion rates. And none of us in that original small circle of analysts really understood the wildlife potential of the program that is now perhaps its hallmark and main source of political staying power, focused as we were on putting conservation on the most erosion-prone land.

No, the real question about the CRP is whether we could do something as significant in agricultural conservation again. Can the analytical case, lobbying skill and grassroots capacity be marshaled again, with the same effectiveness that was on display in the early 1980s around the Conservation Reserve Program, to make the next generation of large-scale improvements to agriculture and conservation policy?

I wish I had the answer. But every time I see a CRP field like the ones I saw in Montana those many years ago, or perhaps the one James Hamilton describes in his estimable book, it renews my determination to try to make something just as amazing happen again. Goodness knows we need it.

Ken Cook

Ken Cook is president and co-founder of Environmental Working Group, a nonprofit research and advocacy organization whose mission is to use the power of public information to protect public health and the environment. Cook has worked at the nexis of agriculture and environmental policy for over 30 years.

INTRODUCTION

Pollution can be easy to spot, but conservation sometimes requires a roadmap. If you drive about 10 miles outside the town of Silk Hope, NC, you may easily miss James Grass's farm. A small strip of trees along the roadside shields from view the grassy fields, growing trees, rural farm house, and well-stocked catfish pond. Mr. Grass has strong attachments to this land. He was born nearly two miles down the road and moved to this farm in 1963. He worked construction, repaired machinery, and raised chickens and cows on the farm. He also raised two daughters there. The connection he feels to this land is deep.[1]

If you paid taxes in 2003 or later in the United States, you also have a connection to this land. Of the 39 acres on the farm, 30 are enrolled in the Conservation Reserve Program (CRP). This program, the largest conservation program in the United

States, pays farmers not to farm their land. Instead, they may be required to leave the land fallow or plant trees to reduce erosion and improve wildlife habitats. In October 2008, there were more than 33 million acres enrolled in the CRP, spread across 46 states. If you drive down a byway or a highway, the chances are in rural America that you have driven through a vista or expanse whose beauty is partly paid for by this conservation program. For reasons explored in this book, the CRP often does not attract wide attention. But, if you stop and ask a farmer how he or she views this government program, you will often hear an interesting tale.

James Grass is happy to talk about how he came to enroll his land in the Conservation Reserve Program. This is actually one of several government programs he has participated in. Remembering when he first learned about the CRP, he said, "I quit growing chickens in 1996 and started attending some local agricultural meetings. In 1999, I had done a conservation program to fence cows out of my streams. I guess I heard about it through the Soil Conservation District of Chatham County." He also noted that he "got a cost share on noncommercial thinning to help prevent pine beetles. I think they paid 70%." Through these earlier involvements with county level agriculture and conservation officials, he found out about the CRP.

Grass explains that he dealt with three different agencies in his decision to place the land in the CRP. Describing the road to enrollment, he said, "First, I went through the Farm Service Agency…The next stop was to get the Forest Service and they came and looked it over and they wrote a plan… and it wasn't our local forest service. They were from an office about 90 miles away in the next county." He added, "The Forest Service are the police of this outfit…but I had a good relationship with the Forest Service ranger." Three trips to the Farm Service Agency (FSA) yielded a contract to place his acres in the Conservation Reserve Program. Paperwork with the county Soil and Conservation Department was also involved.

His interactions with the regulators in the CRP were mixed. He notes that "some people who write the program never see how it is implemented on the ground," and that at least one FSA official involved in the conservation contract on his land "couldn't tell one tree from another." Yet others did know what they were doing, such as the Forest Service worker. Grass says of her, "I appreciate her. She was really doing her job."

There is an elaborate ranking scale to determine which lands the government will pay to enroll in the CRP. In the general CRP signup process, each parcel of land is scored using a measure called the Environmental Benefits Index (EBI), so that government officials can get a relative sense of which lands will deliver bigger environmental gains if they are held back from farming. Some farmers today can use detailed information and geographical information system (GIS) technology to determine how to shape their proposed conservation parcel to score high on the EBI. For Mr. Grass, this process was partly a "black box"—completely opaque. Asked if he had heard about the EBI used to evaluate most CRP lands, he simply said, "No." He guessed that his land was enrolled by the county office of FSA, saying "I think it was eligible because it was up the Jordan Lake watershed."

Grass ended up deciding to take his 30 acres out of farming and agreed to a conservation plan to plant trees on this property. CRP contracts usually run 10–15 years, although for some types of land in North Carolina longer rental agreements are an option. Explaining why he signed a contract to enroll the land for 30 years in a conservation program, he said, "I could get double the money. The money per acre is double the...rate if you do a 30-year contract." Once the land was accepted, he was given a list of government-approved contractors to plant the fields, including a planting of pine trees. Although Grass wanted to have the planting done by hand, the government plan involved machine planting and chemical spraying for site preparation. Grass believes the government's method of tree planting was in part responsible for the line of dead pine trees on the edge of his CRP field.

The CRP plan also called for Grass to plant 10% of the acres with hardwood trees. Yet, he notes that the proper site preparation was not done, so that the grass simply choked out the seedlings. The result was a stretch of land without trees. Looking at the treeless area, he explained, "I had to put 10% in hardwoods, but it is a disaster. But they [FSA] keep saying that I have a healthy stand of trees." He remarked, "The government done what they call 'site preparation,' but they won't do that any more...They did pay for the first year, but they wouldn't come back, so I lost a lot of trees." Although he has shown an FSA official how bad the planting turned out, they have refused to come back and do more.

The local county regulators know he has complied with his contract, but he is "afraid the guys will change. You know, that another person will come [who will not understand why an area slated for trees looks so empty]." Thinking about how to handle this, he is considering trading hunting rights on his CRP land to a lawyer who in turn can provide legal advice and a letter of compliance.

Despite these glitches, much of Grass's 30 acres looks like an advertisement for the success of the CRP. There are rows of trees 5–10 feet in height that have grown during the three years the land as been in the CRP. Surveying the area, Grass noted, "We have quite a lot of wildlife. Deer, cattle, and now groundhogs, but those are not so good." When asked if there is more wildlife now that the acres are planted in trees brought by the CRP plan, he is honestly unsure, saying, "Oh, I don't know. I can't tell the difference." The land is home to two wooden hunting stands that rise about 25 off the ground, used by hunters during deer season.

The 30 acres enrolled in the CRP yield a stream of income for Mr. Grass, amounting in 2003–2005 to over $4,000. When asked why he decided to enter the land into the conservation program, he noted,

> My cows are old and my fences are high and if I have trees [instead of cows on the land], I don't have to worry about feeding them…When we started on it, my wife was 68 and I was 70 and I didn't want to sell. I am a conservationist too, but I don't want to give it away. There are some people who give their land away for conservation. They hate development. I don't hate development. I like timber. I am a timber man. That is one reason I moved here…I like my land. I didn't want to sell…I wanted to retire and keep my land.

The CRP allows James Grass to earn a return from his land for minimal work. He had to mow it during the first year in the CRP and maintain a driving path around it. He says that he has to follow "best management practices…but that might just mean that I have to thin it in 20 years and that will get me more money." Grass notes that whoever owns the land after the 30 years of the conservation contract will have a nice stand of timber to sell. Looking over the acres that provide a bit of retirement income and a possible inheritance for his children, he has a farmer's pride in the growing trees and frustration in the "crop" that failed. As he says, "I am very proud of it—except for that stand of trees that won't grow."

If you took an aerial snapshot of the CRP land, James Grass's farm might be a single point in the image. His farm is one of 5,592 participating farms across the state of North Carolina. Across the United States, there were a total of 417,016 farms that placed 33,573,342 acres in the CRP as of October 2008.[2]

The story of one farmer cannot tell the tale of a regulatory program. Yet the experiences of James Grass do reflect the challenges often involved in living out the high ideals expressed when legislation is first passed. The text of a bill in Congress often identifies a problem and a general description of a solution. It remains for a regulatory body to spell out what the exact terms of that solution will be. A set of regulators in the field in turn makes decisions about what the rules mean on the ground. The many participants in a program try to understand what is required by a rule and how they can comply. This long causal chain that starts with a law and a rule ends up with a set of outcomes in the world, such as streams that are or are not polluted by runoff and habitats that are or are not protected from development.

This book examines the development of a single regulatory program, the Conservation Reserve Program. The primary focus is on the role that information plays throughout the policy cycle. As the title implies, I concentrate on how the program is able to run (often well) on imperfect information. The book is not an attempt to provide a detailed cost-benefit analysis, an examination of how the program might have worked under alternative regulatory structures, or a comparison of how information provision works in agricultural policy versus other government programs. *Conserving Data* is rather an attempt to derive lessons about the role of information in policy development and implementation by examining closely the story of one regulatory program. This case study approach does not allow direct hypothesis testing, but it provides the detailed description necessary to generate hypotheses or lessons that can then be tested in analyzing other programs.

In Chapter 1, I examine how the problem of hidden action and information affects the operation of many relationships in regulatory politics—between voters and legislators, legislators and regulators, and regulators and the public. I then focus on key decision points in the evolution of the CRP and discuss how the design of institutions and problems with information provision affected the operation of the program. Chapter 2 lays out how regulators designed the EBI, the metric used to rank and evaluate which lands should be brought into the CRP. Chapter 3 discusses how the EBI was implemented in the field and shows how the efforts to provide guidance to regulators in the field worked. There are many parties looking over the shoulders of the regulators, and Chapter 4 discusses the picture of the CRP that emerges from three venues: reports from the Government Accountability Office (GAO),[3] oversight hearings in Congress, and the rulemaking decisions made transparent through the notice-and-comment process in the *Federal Register*.

While government bodies develop information to monitor how a regulation is working, the revision and renewal of legislation offer interest groups the chance to produce their own assessments of what is happening in the field. Chapter 5 describes how one interest group, the Environmental Working Group, created and posted a publicly accessible database on the internet that allowed people to track the distribution of farm payments (including CRP funding) and produced studies that influenced the terms and course of debate about farm legislation. Chapter 6 examines how the media covered the evolution of the CRP and how academics assessed its operation. Chapter 7 offers conclusions about how regulations are made and how information provision affects decisions made throughout the policy cycle, based on lessons from the Conservation Reserve Program.

CHAPTER 1

INFORMATION THROUGH THE POLICY CYCLE

When the Conservation Reserve Program shifts cropland out of production, society loses something of value. The crops that once grew on that land are no longer brought to market. In the early days of the program, this was actually one of the reasons for supporting the CRP. With supplies high and commodity prices low, legislators viewed paying farmers not to produce as a way to maintain farm incomes and reduce the supply that generated low prices on the market. When commodity prices eventually rose high enough, however, legislators tended to shift the discussion about the CRP to its environmental benefits. Three prime benefits of removing land from cultivation and planting it with grass cover or trees were often cited: 1) reduced pollution of nearby water sources by runoff from the fields, 2) enhanced benefits from wildlife inhabiting the newly fallow fields, and 3) the reduction in soil erosion and the accompanying long-term gains in soil quality.

Each of these benefits involves a market failure related to information. Consider first the problem posed by chemicals and soils that wash from cropland into nearby water sources. This nonpoint water pollution could be avoided if the world existed as economist Ronald Coase described it. In "The Problem of Social Cost," Coase (1960) asked people to imagine a world with no transaction costs. In this state, information would be free. Negotiations would be costless. People could seamlessly and effortlessly come to agreements about payments for harms and transfers of property rights. Coase pointed out that the current world had negative spillovers from private activities whose values were not fully reflected in market prices—spillovers that economists call negative externalities. Yet, if property rights were fully defined and transaction costs were zero, people harmed by externalities, such as pollution, could bargain their way to efficiency.

In the case of polluting runoff from a field planted with crops, it may be that the farmer has the right to use chemicals to enhance productivity in his fields and the right to allow runoff from these fields to drift into a nearby stream. In this case, a neighbor harmed by the pollution in the stream could bargain with the farmer, paying him to reduce the amount of runoff up to the point that the neighbor's willingness to pay to avoid one more unit of pollution runoff equaled the farmer's cost of averting that additional (i.e., marginal) unit of pollution. If there were multiple neighbors, then in a world of perfect information about damages to the environment and costless negotiations, the farmer could receive many micropayments from neighbors to reduce the pollution. Coase also points out that if the property rights started with the right of the neighbors to be free of pollution, then the farmer might bargain with them for the right to release runoff. In either case, the same amount of runoff would eventually be negotiated.

The wildlife benefits from the increased advantages of land in conservation are examples of positive spillovers (i.e., positive externalities). A farmer might gain pleasure from knowing that more deer and birds inhabit his land. He might hunt the land himself or charge others for the right to hunt. Beyond these values gained from using the land in conservation, there is also an existence value that some people might place on the knowledge that flocks of birds and colonies of rabbits have richer and deeper habitats. In the alternative world that Coase invited people to consider, those with existence values for wildlife could pay a farmer to leave more fields planted with trees and cover grasses. The negotiations could continue until the marginal cost to the farmer of the conservation of an additional unit of land was equal to the marginal benefit—including the habitat benefit derived from the way other people value the conserved land and animals left free to roam it.

Erosion itself involves a third type of market failure rooted in information. When a farmer leaves a field fallow for a period of time, this may increase the long-run productivity of the soil. If markets functioned perfectly and information were available without cost, the farmer would be rewarded for greater long-term productivity with higher land prices for the fallow field. Yet, if information were costly, the farmer or the potential buyers might not see the full benefits of leaving an additional section of land in conservation. Or the farmer might have a shorter time horizon than the long-term view adopted by social planners. In this case, the producer might jump at short-term gains in crop production, even if it meant more soil erosion and less production in the long run. Separating out the exact contribution of the lack of information about the value of conservation versus the difference in the time horizon between a farmer and society is difficult, considering the problems inherent in how a farmer may factor soil productivity into his decisions.

Coase won the Nobel Prize in part for his analysis of the types of problems posed by externalities, such as pollution and wildlife conservation. He lamented that his article on externalities was frequently cited, but often misunderstood. Many analysts transformed his "what if" to a "since," meaning that they minimized the problems associated with negative spillovers by pointing out how proper definition of property rights, information provision, and negotiation could lead to bargained resolutions that could lead to efficient amounts of pollution or wildlife. Coase himself stressed that he posed the thought experiment of a world with free information and costless bargaining as a counterfactual case, meant to cause one to pause and think why the violation of these assumptions leads to so many different types of market failures.

In the case of cropland runoff containing chemicals and soil that reduce water quality, markets are clearly imperfect. When confronted with a polluted water source, people generally do not know where the runoff comes from, what damages arise from the chemicals and soil, and whom to bargain with to reduce this problem. The logic of collective action also leads to free riding—letting someone else worry about solving the problem. As for the value humans place on knowing that birds and other wildlife have better habitats, putting a dollar amount on it is difficult, just as collecting fees to transfer to producers to put land into conservation may be difficult. Individuals again have the incentive to sit back and let others work to maintain and restore habitats.

After the Farm Bill of 1990, the versions of the CRP that emerged were seen as partial solutions to these market failures. Involving government in paying producers to reduce pollution and increase habitat does help solve the collective action problem. Rather than trying to convince those who benefit from lands in conservation (in the CRP) to pay voluntary transfers, the government uses general revenue funds from taxpayers to pay the rental rates in the CRP. The lack of information about the nature of the lands and exact benefits from conservation, however, mean that there is no guarantee that the operation of the CRP is optimal. When one adds in the large number of people involved and the many decisions entailed, the management problems posed by the CRP are daunting. In one sense, the CRP demonstrates an attempt to remedy imperfect markets through the institutions of imperfect government. This chapter explores in detail the singular role that information plays in the development and implementation of government regulations. It also looks at the typical design of regulatory institutions and then explains the particular architecture of the CRP.

PRINCIPALS AND AGENTS, VOTERS AND LEGISLATORS

In one sense, government institutions are a series of principal-agent relationships, a view set forth by Kiewiet and McCubbins (1991) in *The Logic of Delegation.*[1] Principals are people who delegate decisionmaking power to others, namely, their agents. There are great gains from this arrangement. The division of labor allows the principals to pursue other interests and agents to focus on specific tasks and gain expertise and experience. Since information is not free, principal-agent relationships carry potential problems. Hidden action is a danger. Agents may make decisions different than those that principals intend, but the actions of agents may not be fully visible. (Agents may also opt for on-the-job leisure consumption—the language economists use to describe the problem of agents not expending enough effort at work).

A second problem lies in hidden information. A principal may not know all the options available to an agent or what information the agent had when making a choice. The prospect of hidden action and hidden information sets a potential stage for agents to pursue actions at variance with the principals' choices, if they were fully informed and making the decisions. Agents with multiple principals face an added difficulty of deciding how to follow the directions of their principals, if the principals disagree among themselves about what course to pursue.

One approach that principals take to deal with agency problems is to set up institutional constraints. Principals can design contracts that try to align the interest of agents with their own. Sorting and screening is a second option, where principals devote a great deal of time to considering up front which people to hire as agents. A third method is to monitor agents and establish reporting requirements. In a way, this gives back some of the gains of delegation. A principal hires an agent so that the principal does not have to worry about a set of decisions and their implementation. The principal then devotes a portion of agency resources to get reports on what agents are doing and to investigate sets of agent decisions. A final way of dealing with agency problems is to delegate decisions to multiple agents, who then compete to serve the principal's interests. The actions of one group of agents then demonstrate what is possible and serve to spur the activities of other agents.

The relationship between voters and members of Congress is a complex principal-agent relationship. Single voters implicitly delegate thousands of decisions to their district's U.S. Representative and Senator. These legislators will cast hundreds of votes of interest to the constituent over the two-year term. Douglas Arnold (1992), in *The Logic of Congressional Action*, lays out the chain of reasoning that may affect how members of Congress consider voting on a piece of legislation. If the legislators are primarily focused on re-election, they may calculate the likelihood that a particular vote will end up being considered by a constituent voting in November. The benefits and costs of the bill affect this likelihood as well. If the bill involves benefits or costs that are large in magnitude for individual voters, they may be more likely to notice a vote in Congress. If the costs and benefits affect the local area or an interest group a voter is a part of, the likelihood of recognition also goes up.

For many issues, there is a long causal chain between what goes on in Congress and what happens in the real world as a result of policy changes.

If the causal chain is relatively short, then a voter may recognize the link between change and a vote. If the income tax rate is cut for married couples, for example, that change is easy to recognize. If one assumes that higher pay for public school teachers increases teacher retention, which increases student performance, which attracts employers looking for an educated workforce to a state, then the voter may not link new jobs from a new plant opening to higher teacher pay. Voters may be more likely to see the causal chain if they know someone directly affected by a bill, for example, an oil industry worker who is drilling in a new area opened by legislation.

A member of Congress may realize that constituents will focus on legislation down the road if there is an instigator in the district willing to spend time and money to advertise how the legislator voted on it. Instigators can be challengers from another party or interest groups willing to develop and buy ads to defeat the incumbent. Finally, the legislator may have to reason whether a voter will have a forward-looking perspective and judge the candidate or party's position on issues or, alternatively, whether they may look retrospectively for direction, at how a party or an incumbent handled decisions in the past. The ability of candidates to frame elections about the past/future and person/party provides some leeway for members of Congress to try and influence the decision rule (forward or retrospective) a constituent will bring to the voting booth.

The images and ideas a voter carries in his head when voting depend in part on the operation of information markets. In *An Economic Theory of Democracy*, Anthony Downs (1957) outlined the four major types of information people demand: consumer, producer, entertainment, and voter. Markets for the first three types of information work fairly well. If people do not seek out information, they do not gain the benefit. This means that consumers, for example, invest in learning about cars they might buy, restaurants they might visit, and movies they might see. Producers, however, seek out data that help them make better decisions at work. Entertainment information includes things people like to consume simply for their own enjoyment. The value is intrinsic; for example, a person may simply like to read a magazine, tune into a particular television program, or listen to music podcasts.

The logic is different in the operation of the market for civic or voter information. It may be that a voter values the policies and decisionmaking capabilities of one candidate over another candidate. More information about the candidates might increase the probability that the voter will select the correct candidate, based on the voter's underlying preferences. However, the probability that one person's vote will be decisive is so low in a congressional race that the expected benefits of getting information about a congressional candidate are tiny. The opportunity costs of becoming informed are real, in terms of time spent reading or subscription fees. This means that for most people the net return for learning about politics is negative. Downs called this "rational ignorance." From an individual's standpoint, not learning the details of how a member of Congress votes is ideal. If a constituent's vote does not matter in a statistical sense, why bother gathering more information?

There are obviously segments of society that do demand hard news. Those who feel a duty to vote may also feel a duty to stay informed about politics. For some individuals, the details of policies and politics are inherently interesting. If media outlets choose to cover politics as a horserace or an arena for human-interest tales, then the entertainment angle will draw readers and viewers to political questions. Overall, however, most people will remain rationally ignorant about the details of policies.

This means that many law and regulation stories will go unwritten and untold if a media outlet is focused on audience and profits. If media revenue comes primarily from selling the attention of audiences to advertisers or gaining subscriptions from interested readers, then for-profit media outlets will avoid many types of policy stories. In the broadcast television realm, deregulation and increasing competition from the entertainment offerings on cable mean that major votes in Congress are much less likely to be discussed on the network evening news now.[2]

Information is usually created with incentives in mind. The commercial media hope to translate viewers' (or listeners' or readers') attention into sales for advertisers or hope to generate subscriptions. Nonprofit media hope to change the set of ideas or issues that audiences think about. Another group trying to reach people, however, does not try to change the public's purchasing decisions. They hope to change people's votes. Incumbents in Congress pay keen attention (and a significant amount of time) to raising the money that allows them to communicate with their main principals, the voters. In a world where attention is diffuse and news media are focused on audience interest, campaigns try to produce their own information to vie for voters' attention. House members running for reelection in 2006 on average raised

$1.3 million, while senators running for reelection raised an average of $9.5 million.[3] By the end of their campaigns, these candidates often attempted to translate these dollars into votes through expenditures on information. Ads bought on television and radio, and in print and electronic media, and letters and cards sent through the mail offer a campaign the chance to reach voters with broad and targeted messages.

In the three-year period 2005–2007, House members gathered over $700 million in campaign contributions.[4] Seventy-nine percent of this money came from donors outside U.S. Representatives' own districts. In effect, the state of media markets and campaign finance laws mean that the principal-agent monitoring within a congressional district depends heavily on money from outside the area. Members of Congress voting on legislation, particularly bills valued by special interests, need to trade off many factors: the likelihood that a vote will matter in their districts; the need to raise money outside their districts to fund campaign expenditures; and potential conflict, if any, between what generates contributions outside a district and what generates attention within a district.

REGULATING THE REGULATORS

A second principal-agent relationship arises when legislators delegate decisionmaking powers to a regulatory agency. The way that Congress writes the text of legislation affects the degree of delegation. If legislators do not trust the agency (perhaps because the top officials were appointed by a president from the opposite party), then they may write very specific language into a measure and try to get that text through the legislative process. If members of Congress trust an agency and are willing to defer to its expertise, they may at times explicitly delegate particular questions or definitions to the regulatory body to determine after a law is passed.

Once the power is shifted to the regulatory body, Congress can choose to monitor decisions by regulators in at least two different ways (as described by McCubbins and Schwartz in their classic 1984 article, "Congressional Oversight Overlooked: Police Patrols versus Fire Alarms"). An example of police patrol monitoring is when members of Congress decide to examine a subset of an agency's decisions. This may entail congressional oversight hearings, where agency officials must answer questions about the operation of a program. Congress might also delegate investigations to the GAO, requiring it to inspect how a program is being implemented. The prospect of police patrol regulation may induce an agency more often to make the decisions that congressional principals desire. The agency may face difficulties, however, since there may be multiple sets of principals in Congress, sending it conflicting signals about policy. Different congressional committees, different congressional leaders, and legislators from different regions may all have different policy positions they wish to see the agency embrace.

An alternative method of monitoring is the fire alarm process. In this method, Congress waits for someone—constituents, interest groups, other government agencies—to "pull the fire alarm" and complain about the implementation of a policy. There are distinct advantages to this process. Citizens and interest groups bear the costs of monitoring and collecting the information that draws attention to a policy. Congress does not have to hold regular oversight hearings, which U.S. Representatives and Senators view as wasted effort if they primarily involve monitoring decisions that turn out to be uncontroversial or inconsequential. In contrast, the police patrol option allows members of Congress to act and investigate an agency's choices in the areas that matter to constituents and interest groups.

The agency does have some advantages if officials choose to pursue directions at odds with the original goals expressed in the legislation. They can take actions that remain hidden from congressional oversight or view (at least for a time), and they may possess information that is unavailable to members of Congress who are less expert in the policy area. If the agency takes actions against the interests of its congressional principals, it may be politically costly for Congress to pass new legislation that overturns the agency action. Passing new legislation might reveal embarrassing information to constituents— for example, cause voters to wonder why legislators allowed an agency to stray so far from an intended position.

One way that members of Congress can constrain the actions of regulators is through the rules that govern the rulemaking process. McCubbins, Noll, and Weingast (1987) highlighted this point in the title of their article, "Administrative Procedures as Instruments of Political Control." Under the provisions of the Administrative Procedure Act (APA), most regulations follow a standard path as they are formulated. A notice of proposed rulemaking is announced by an agency in the *Federal Register*. The agency next circulates the text of the proposed rule in the *Federal Register* and solicits comments. Interest groups, individuals, and other government bodies will often file comments with the agency, providing studies and information about their reactions to the proposed rule. The agency takes this feedback into account and publishes the final text of the rule in the *Federal Register*. At this time, the agency must indicate how it decided to respond to the comments it received. If the agency simply ignores substantive comments, parties who disagree with the rule can attempt to overturn it in court on the grounds that the agency acted in an "arbitrary and capricious" manner. In this way, the comments and ideas provided during rulemaking may serve as a constraint on agency action.

The public, transparent nature of notice-and-comment rulemaking carries many benefits for congressional principals. As McCubbins, Noll, and Weingast (1987) pointed out, it rules out surprise regulations. An agency cannot simply declare new policy overnight. The process allows a member of Congress's favored constituency to present its own reactions to potential regulatory decisions and provide valuable advice to policymakers. The extensive time required to develop a rule allows legislators to give feedback to the agency before a rule is finalized. The text of a proposed rule can serve as a political trial balloon, allowing regulators to see how interest groups and constituents will react to a potential policy.

The existence of the rulemaking stage removes some pressure on legislators as they draft bills. They do not need to determine all definitions and options in a policy area because they can choose to delegate some decisions to the agency, which determines them during rulemaking. The public comment process gives those interest groups that a member of Congress cares about a chance to participate in the final policy selection, since the studies and comments submitted may provide helpful information and constrain agency behavior. As legislators craft a particular bill, they can even draft in specific procedures that will affect rulemaking on that measure. They can, for example, indicate whether costs can be taken into account during the formulation of a rule or require that a particular study be performed as rulemaking or implementation proceeds.

This description of the rulemaking process makes it appear as if congressional leaders are strategic and regulators are passive in complying with monitoring procedures. Regulators may be strategic, however, in their decisions about how to issue a rule. In some circumstances, an agency may decide to avoid the notice-and-comment rulemaking process to issue a rule. Instead, it may signal to the regulated community what it expects in terms of actions and behavior by issuing guidance documents and policy memos. The degree that an agency can use these informal rules depends on the leverage that it has with a regulated party. If the agency tries to do too much through informal rules, the regulated party might challenge the approach in court by claiming that the agency is trying to set new policy through memos. Yet, if the regulated party depends on the discretion of the agency in future interactions—for example, it needs permits in the future or expects to interact with agency officials over time in enforcement situations—then the regulated party may not challenge the legality of an informal rule. The growth of informal rules means that some policy memos and guidance documents are now published in the *Federal Register* as a way of providing information to the regulated community.

Note that the terms formal rulemaking and informal rulemaking, which are common in the political science literature, have different meanings in the APA. Per this law, formal rulemaking describes the use of a trial-like procedure before an administrative law judge to resolve issues, such as rate-making cases. Also under the APA, the term informal rulemaking refers to the notice-and-comment process. This book employs different definitions. Here, formal rulemaking means the notice-and-comment rulemaking process plus the many procedural requirements that surround this way of developing a rule. Informal rulemaking, in this analysis of the CRP, includes the guidance documents and policy memos issued by an agency because through them an agency can convey expectations about what it wants regulated parties to do and sidestep the more costly publication and comment process.

In "Strategic Regulators and the Choice of Rulemaking Procedures", Hamilton and Schroeder (1994) outlined when one might expect an agency to use informal rulemaking rather than taking the notice-and-comment rule-making route. They reasoned that the higher the transaction costs of securing an agreement, the political costs of pursuing the notice-and-comment route, or the regulatory costs imposed on parties, the more likely an agency might try to use guidance documents and policy memos. If Congress is extremely specific in the wording of the text of legislation, then regulators can use informal rules to avoid the constraints they would face in more formal rulemaking. If an agency is concerned with uniformity, enforceability, and the value of precedent, regulators may choose the formal rulemaking route and publish in the *Federal Register*. Note that adopting a rule through the notice-and-comment process can make standards more widely recognized and clearer, which can reduce the pressure on regulators for individual negotiations and decisions about what is actually required by a rule.

WATCHING THE FIELD

Once a rule is formally enunciated by publication in the *Federal Register* or informally announced through circulation as a guidance document or policy memo, another round of delegation ensues. Regulators in Washington, DC, depend on agency officials spread across the United States to interpret and enforce rules. These on-the-ground regulators, in turn, face problems unanticipated by or unaddressed by the legislators. The actual effect of a rule—the likelihood that it will change behavior in the field—may depend on how local regulatory officials act. These local regulators may be in state or regional offices of federal agencies or, in cases where federal implementation is delegated to other entities, they may be officials from state agencies.

A final, implicit principal-agent relationship arises between regulators and those whose actions they monitor. At times, regulators on the ground may be principals, choosing to delegate particular decisions to the regulated community. Officials may rely on regulated parties to provide information about the operation of a plant or facility and self-report on compliance. The degree that producers actually comply with regulations may in part depend on how long a requirement has been in operation. In the early stages of a regulation, regulated parties may fail to comply because they are ignorant of its existence or unsure of its meaning. As information campaigns, enforcement actions, interactions with regulators, and discussions at industry gatherings unfold, noncompliance from ignorance declines over time. Some parties may still avoid following a rule, based on strategic calculations of the likelihood they can evade detection or a fine. Across some areas of regulation, compliance is relatively high even in the absence of intense inspections or high fines, with theories for this result ranging from the importance of reputation effects in some consumer markets to the importance of ethical norms in predicting compliance with the law.[5]

Policymakers in Washington can learn how rules are implemented in the field via at least three avenues. One, police patrol monitoring, entails sending agency investigators to audit samples of compliance with a rule. Congress may prefer to generate its own information by requiring GAO to investigate a regulatory program and produce a separate report on the agency. Regular, intensive oversight hearings by congressional committees help members of Congress sample what is going on in a program. Two, legislators can learn about implementation through fire alarm monitoring. Agency hotlines allow people to complain (often anonymously) about regulatory decisions and compliance. Constituents or interest groups may offer feedback directly to members of Congress about a rule's provisions and even take the added step of producing their own reports and evidence on the matter. Three, court cases serve as a source of information about field operations. If a regulated party strongly disagrees with a rule's interpretation, the individual or company or interest group can take the agency to court. Unpopular rules can be attacked strategically in court cases if an interest group can prove that the rule was issued through improper procedures (the aforementioned "arbitrary and capricious" manner). Cases about regulatory compliance also show Congress how rules are being translated into requirements in the field.

INFORMATION AND THE CONSERVATION RESERVE PROGRAM

The rules that define the operation of the CRP originate in omnibus farm legislation that passes in Congress every five to six years. The farm bill usually has highly concentrated benefits (the commodity subsidies that flow to agricultural producers) and widely dispersed costs (the set of tax revenues that fund this redistribution). Farm legislation often attracts focused, intense lobbying and generous political contributions. In the 2006 election cycle, agribusiness interests contributed nearly $45 million to federal candidates and party committees.[6] Crop growers and processors contributed $12.7 million of these funds, while food manufacturers and retail food stores accounted for

$7.8 million. In addition to campaign contributions, the agribusiness sector spent over $91 million lobbying Congress and federal agencies in 2006. The political battles in farm policy are often fought along regional lines, with representatives from farm districts supporting their commodity producers and legislators from urban areas asking why subsidies should flow to farming families with relatively high net assets.

In each farm bill, the CRP brings in additional members to the coalition supporting the bill. The CRP directs payments to more producers than many other subsidy programs. As of October 2008, for example, USDA had 739,160 active CRP contracts spread across 417,016 farms.[7] The increase in wildlife habitats on land withdrawn from crop production has attracted the support of hunters for this agricultural program. Representatives of Ducks Unlimited and Pheasants Forever regularly voice support of the CRP during farm bill debates. The water quality and wildlife habitat provisions of the CRP attract the support of environmental groups, including the Environmental Defense Fund, Environmental Working Group, and Sierra Club.

Most voters in the United States remain rationally ignorant about the provisions of the farm bill and the positions of their legislators. Media coverage of the debates over the CRP is clustered in farm-area newspapers and periodicals, and outlets specifically targeted at farmers, environmentalists, or hunters. The actual text of the legislation dealing with the CRP is sparse, but may set the number of acres that can be brought into the program and lay out the broad goals of the program: protection of water quality, enhancement of wildlife habitats, and prevention of soil erosion.

The lack of specific requirements spelled out in the legislation governing the CRP gives FSA great freedom in deciding how to bring lands into conservation. FSA developed a scoring system, the EBI, to rate croplands in different environmental dimensions. To enter the CRP, producers (i.e., the landowners, predominantly farmers) now must submit bids that specify the characteristics of the land, the conservation practices they will pursue, and the rental rate they will accept. After producers submit their bids and the EBI scores are determined, the secretary of agriculture specifies a cutoff EBI number. Lands with higher scores than that figure can be brought in to the CRP, and the government pays producers not to plant crops on those parcels of land for 10 to 15 years.

While the overall operation of the CRP has been the subject of notice-and-comment rulemaking in the *Federal Register*, FSA did not develop and issue the EBI through this procedure. Instead, it developed the EBI in Washington. At first the formula was kept secret, even from agents in the field. In 1996 a transparent EBI was released, so that producers could see how the measure was calculated and alter their conservation proposals to increase the likelihood that their land would be accepted. The EBI scoring system was published in agency handbooks and directives, rather than the *Federal Register*. This gave FSA the freedom to alter the questions and weights in the EBI over time. Each signup round for the CRP allows the agency to add new criteria and drop questions that are hard to score or prone to errors in calculation.

One reason that FSA is able to avoid the more costly notice and comment process to develop and refine the EBI is that many people do not view the CRP as a regulatory program. The agency does not use command-and-control power to promote conservation; it does not tell a particular producer to take particular croplands out of production or plant specific trees. Farmer participation in the CRP is voluntary. The government, in effect, is saying that participation in this program—where the government pays rent for the land—means accepting that certain definitions of property rights apply and certain rules must be followed. The theory of loss aversion suggests that while people will strongly resist policies that impose a loss from the status quo, the prospects of gains may not set off disputes as intense over the types of gains to be awarded.[8] The flexibility of not using the *Federal Register* to define the EBI means that FSA can use a system that weights particular dimensions of environmental protection more highly, without having to defend the calculation process in formal rulemaking.

The legal basis for FSA to avoid using the formal rulemaking process, which might require publishing frequent draft changes of the EBI in the *Federal Register* and asking for feedback, lies in the APA. In 5 U.S.C. §553(a)(2) of the APA, the language exempts matters "relating to agency management or personnel or to public property, loans, grants, benefits, or contracts," from notice-and-comment rulemaking requirements.[9] Since the EBI relates to the contracting process in the CRP, this may allow the agency to keep the detailed development and changes to the EBI out of the formal rulemaking process.

The freedom of FSA to develop the EBI does not mean that it has unlimited discretion in the implementation of the CRP. In fact, there are both police patrols and fire alarms built into the operation of the program. USDA's Office of Inspector General (OIG) has conducted audits of CRP contracts. This same office has a regulatory hotline that people can call to complain about the operation of the program. Congress intermittently holds oversight hearings on CRP issues, including the EBI. Legislators also task GAO with investigating how the CRP works in the field and how it can be improved. With each renewal of the omnibus farm legislation, media coverage and interest group studies focus attention on the way the CRP operates. As technology improves and data become more widely available, academic studies have started to examine the impact the CRP has had on the actual environmental benefits delivered by lands brought into the CRP.

In the world of zero transaction costs proposed as a thought experiment by Ronald Coase, checking on the operation and enforcement of regulations in the CRP would be effortless. One could see and evaluate the conservation benefits generated by each parcel of land proposed for the CRP and observe how FSA made decisions about which parcels to rent. However, in a world of costly information, effort, and negotiations, determining whether FSA made the right call on each CRP contract would be extremely costly. In October 2008, there were 33,573,342 acres rented in the CRP, with annual payments of $1.7 billion spread across more than 400,000 farms.[10]

The design of congressional and agency monitoring institutions, however, provides a way to monitor the implementation of the CRP without excessive transaction costs. The setup of the many principal-agent relationships involved in this program allows principals in many instances to economize on the costs of information gathering. Not all decisions will need to be vetted or re-examined. Imperfect, periodic police patrols or fire alarms may help regulators in Washington be aware, in part, of what goes on in the field. The following chapters examine how information flows from Washington to regulators and producers in the (literal) fields, and how feedback and changes in technology have improved the operation of parts of the CRP process over time.

Policy analyses often read like mysteries, where implications are presented at the end of the tale. By stating the lessons learned from the CRP up front, with luck the story of the CRP presented in the chapters to come will elicit a greater appreciation for the many different ways that (imperfect) information is generated during the operation of the program.

CHAPTER 2

DEFINING THE ENVIRONMENTAL BENEFITS INDEX

If information were free and ubiquitous, running the Conservation Reserve Program would be a relatively easy job. Farmers could readily identify the nature of their land and the benefits to be gained by transferring their parcels from farming to conservation. Regulators in Washington could costlessly assign dollar values to the environmental benefits associated with each parcel. With a budget provided by Congress, USDA could simply start spending where the benefit-to-cost ratios were the highest and keep buying contracts until the funds were exhausted. This approach would maximize the net benefits to society of placing farmlands in conservation. If property rights to the environment were well defined and information and negotiation were truly free, then there might not even be a need for the CRP. People could costlessly negotiate with farmers to place their lands in conservation, and millions of tiny payments could flow from those who benefit from the conservation efforts to those landowners choosing to forgo agricultural production.

This imaginary world is very different from the actual agricultural policymaking environment in Washington, DC. Regulation in the real world involves costly information and negotiations. Decisions are based on imperfect and imprecise data. Whether a particular set of interests is represented may depend on whether the costs and benefits of a policy change are widely distributed or concentrated, and whether the costs of political organizing are relatively low. This chapter draws from interviews conducted for this book with many different regulators[1] to explore how the government developed a way to solve one information problem—how to prioritize lands for CRP contracts. Today, farmers interested in participating in the conservation program can easily calculate how their parcels will score on the Environmental Benefits Index and whether the parcels qualify for inclusion during the general signup periods for the CRP. Understanding how the EBI came to be defined and how it has changed over time is the focus of this chapter.

1990: THE SWITCH FROM SUPPLY REDUCTION TO CONSERVATION

The Food Securities Act of 1985 established the CRP, which was designed to bring up to 45 million cropland acres into conservation rental agreements by 1990. The CRP was designed to induce farmers take their land out of production and plant it with cover vegetation to achieve seven goals: 1) reduce water erosion, 2) reduce wind erosion, 3) reduce sedimentation, 4) improve water quality, 5) create better habitat for fish and wildlife through improved food and cover, 6) curb production of surplus commodities, and 7) provide needed income to support farmers.[2]

In the original act, farmers applied through their county Agricultural Stabilization and Conservation Service (ASCS) offices to sign up for the CRP. Each farmer was free to name a rental fee for his parcel. If the cropland was considered "highly erodible," and if the rental fee was under the ceiling determined by USDA, then the land could be accepted into the CRP. Farmers did not know the rental fee ceiling when naming their price, which regulators hoped would encourage competition among farmers to enter the program and thus reduce the cost of drawing land into conservation. Over the history of the CRP, the lengths of rental agreements that farmers entered into during general signups for the CRP have normally been 10–15 years.

The initial criteria for determining which lands were "highly erodible," and therefore eligible for inclusion, were spelled out in the final CRP rule published in the *Federal Register* on February 11, 1987. USDA developed a way to measure potential soil erosion, called the Erodibility Index (EI), and explained its construction in this way:

> The EI is an index number which indicates a soil's potential to erode. The higher the index, the greater the potential of the soil to erode and the more difficult it is to control erosion if used for intensive crop production. The EI uses the Universal Soil Loss Equation (USLE) and Wind Erosion Equation (WEQ) factors representing the effect of climate, topography, and soil properties on erosion...Thus, the EI estimates the soil's potential to erode, and disregards the applicant's past use of the land. An EI equal to or greater than 8 was selected to identify highly erodible land.[3]

The calculation of the EI was an early forerunner of the EBI, and this focus on erodibility was a rough way to factor in one aspect of the environmental benefits of taking a land out of crop use. Additional restrictions on CRP participation included a provision that no individual farmer could receive more than $50,000 a year in CRP rental payments. To prevent disruptions, such as a reduction in demand at farm supply outlets, no more than a quarter of a county's total cropland could be enrolled in the CRP at one time (unless the secretary of agriculture determined that a larger proportion of lands being taken out of production would not have adverse effects on the local economy).[4]

Once farmland was enrolled in the CRP, it could not be used for commercial purposes. Farmers had to submit and follow a conservation program, which often entailed planting vegetation on fallow lands to restore nutrients to the soil and reduce erosion. Permissible vegetation included grasses and trees, with the government's preference for trees reflected in a goal of planting 12.5% of CRP lands with trees by 1990. As an incentive to keep the lands planted with trees, even after the rental agreement expired, the farmers would eventually be able to harvest the timber. In return for keeping the land out of crop rotation, the farmer received an annual rental payment and up to half the cost entailed in establishing the vegetation cover.

The large scope of the program is evident in early signups. In fiscal year 1987, 126,421 farms enrolled in the CRP and brought nearly 13.6 million acres into conservation rental agreements; 779,000 of those acres were converted to trees.[5] The average CRP enrollment contract was 109.2 acres.

The initial enrollment years of the CRP occurred when farm policies were aimed primarily at controlling commodity supplies. The 1990 Farm Bill, however, began the transition of the CRP from a commodity control program toward a more environmentally-focused policy. This is evident, in part, in the language of the bill, which provided that, "in determining the acceptability of contract offers, the Secretary may…take into consideration the extent to which enrollment of the land that is the subject of the contract offer would improve soil resources, water quality, wildlife habitat, or provide other environmental benefits."[6]

The increased emphasis in the 1990 Farm Bill on the environmental benefits of lands brought into the CRP was part of the changing coalition of farm legislation. Funding for nutrition programs, such as food stamps, was often included in agriculture bills to win support from urban legislators. As concern for the federal deficit heightened attention in 1990 on the size of farm program payments, more provisions to protect the environment were included in the farm bill as a way to sustain support for agriculture spending. Media coverage of legislation in 1990 referred to the "Greening of the Congressional Consciousness" and pointed out the large number of environmental measures being considered by Congress that year, including bills dealing with clean air, hazardous waste, and such agricultural issues as pesticides and wetlands protection.[7]

Environmental groups lobbied strongly to increase the levels of environmental protection in the 1990 agricultural bill, and were aided in their lobbying by strong public interest in environmental issues. Provisions to protect the environment were included in the bill to gain votes on the floor of the House and Senate during the debate over the bill.[8] Environmentalists believed that incorporating environmental provisions in the CRP was more likely to protect wetlands and water quality than trying to gain support for a new program. One political advantage of the CRP payments authorized by the legislation was that the flow of money would increase the number of farmers with a monetary stake in supporting conservation.

How did a desire to change the type of land eligible for the CRP get translated into decisions in the field? The answer lies in the story of the development of the EBI, a metric used to rank the likely benefits of pulling cropland out of production and developing a conservation program for that land. The government officials involved in the development of the EBI tell a consistent tale about the way the EBI was initially formulated.

Tim Osborn worked on the CRP in USDA's Economic Research Service from 1988 to 1998. Describing the early days of the CRP, he noted that

> an agency within the USDA, the ASCS, now the FSA, had a mandate from the Food Securities Act of 1985 to enroll 40–45 million acres in the program. There were some general goals relating to reducing soil erosion and it [*sic*] may have mentioned water quality. And there was sort of a tree goal. But in the early years, they wanted to meet that 40–45-million acre goal. That was the challenge…What they initially did was set up a set of eligibility requirements that would provide the necessary pool of acres they could draw upon while trying to meet those goals.

Osborn said that the early eligibility criteria were simple: "It had to be cropland and there had to be certain erodibility levels." Most of the erodibility focus was wind erosion. This meant that "a lot of land that was enrolled in the program was out there in the plains…People started to ask is this what we really intended? We could get more benefits if we better targeted this land." Examining how the CRP was being implemented, USDA officials realized that "we didn't target it as well as we could have for the dollars we were spending."

According to Osborn, a single phone call started the chain of conversations that led to his work on changing how lands were ranked in the CRP. He related, "One day I got a call from a guy named Jack Web. He was a guy in the ASCS who ran the CRP. He wanted to talk about the possibility of creating a scoring approach for the subsequent programs [signups]…I went over and talked to him and it was a really interesting program from an economic standpoint." After this discussion of a scoring approach, Osborn explained that "over the next couple of days, I sketched out what would be an approach. We would take what goals were—that we wanted to target and we would have a mathematical formula that would capture the benefits that streamed from that, like reducing on-farm soil productivity loss and water quality...There was also the thought of what happens to the soil when it gets off of the farm." Osborn noted that if a farm is near "a major metro area and is cranking a lot of soil in the air," there is a greater incentive to take that land out of production if the wind were likely to blow the airborne soil particles into the city.

In textbooks, the task of monetizing environmental benefits is simple if prices are known and quantities defined. Yet, in the real world of 1990, the lack of this information meant a ranking method was the preferable way to go. Osborn explained, "We knew it would be great to come up with dollar-denominated benefits, but a scoring method was the second best way. We didn't have a common yardstick between the different factors so they could all come together at the end." Osborn and coworkers shared these ideas about the EBI in agency and interagency meetings. Ultimately, he said that "we had a CRP interagency group...We presented to that group and everyone said it sounded really good. In the subsequent three signups, we used that approach."

The initial drafting of the EBI was not subject to public scrutiny, since the discussions were internal to the agency. As Osborn noted,

> there certainly wasn't any pressure in those first few days when I developed the initial one [draft EBI]. Once we got into a group environment, we talked about relative weights...NRCS [Natural Resources Conservation Service] were concerned about what capabilities existed and what could actually be done. They had a huge soils database that hadn't been used for a whole lot, but when the EBI came around, it became extremely valuable. When Fish and Wildlife Service got involved, they wanted the EBI to be related to wildlife as much as possible. I don't think things were trying to be slanted in their particular direction though.

The other principal architect of the EBI was Ralph Heimlich, who worked on its development in 1990 and participated in its revisions after the 1996 and 2002 Farm Bills. After leaving government service, he served as a consultant to Environmental Defense Fund and helped it develop recommendations for changing the EBI and CRP signup process. Heimlich had a similar view of how the need for the EBI arose.

> After the 1990 Farm Bill, when acreage was capped at 36 million acres, some mechanism had to be found to ration enrollment because there would always be far more land willing to bid into the program than could be accepted.

> USDA was sincerely trying to implement the program as an environmental or conservation program, recognizing that the economic effects of enrolling any possible set of 36 million acres were about the same. That is, just setting the level achieved the economic goals of the program at the national level, but how it was done would have a large impact on the environmental results of the program. FSA (then ASCS) wanted a simple, but defensible, nationally consistent enrollment process to evaluate bids that would consider both the benefits (actually effectiveness or potential effectiveness) against the costs of each parcel. USDA was unusually liberal in considering the input of outside agencies, such as the Fish and Wildlife Service and EPA [Environmental Protection Agency], in devising the scoring, but wanted to retain tight control of the structure of the scoring itself.
>
> Prior work by ERS [Economic Research Service]-USDA had shown that considering the cost-effectiveness of each parcel bid could improve the overall performance of the program, and Osborn and I were in a unique position to advise top leadership in USDA about how to consider cost-effectiveness in a practical and timely manner over thousands of bids submitted in each signup. A more ambitious evaluation proposal would have failed because it could not have been implemented in a timely and economic fashion, but the scheme we developed only needed readily available secondary data, data on the parcel itself, and commonly used measurements (such as sheet and rill and wind erosion), and could be done quickly and cheaply.

Deciding how to value different aspects of a land's contribution to the environment inherently involved judgment calls. In the setup of the EBI, there was not a lot of input from politicians, in part because the process was not occurring in the public view, as it would in a rulemaking. Heimlich noted that, "one area that we *wanted* political appointee input in the initial stages was weighting of the various component scores (wildlife, water quality, soil erosion, air quality, etc.). Political appointees were uniformly uninterested in making these choices (which are essentially political and not technical). Instead, they wanted to see a number of weighting schemes and were content to choose equal weighting based primarily on the geographic incidence of enrollment expected with that weighting scheme."

Heimlich reported that once the EBI was included, the public and politicians became more concerned about the geographic incidence of lands that were accepted into the program based on the EBI formula. "After the EBI was used in a few signups [10th and 11th], there was near-constant political and interest group pressure to make any number of changes…Rep. Helen Chenoweth (R-ID) was particularly active in questioning the effects of this [air quality and wind erosion] scoring on potential participants in her district." Looking at the first design of the EBI, Heimlich said that, overall, "the most controversial aspect of the initial design of the EBI was an attempt to use more disaggregated wind-erosion estimates, population weighting, and prevailing wind direction for the air quality component of the score. This tended to favor areas nearer urban centers, particularly in interior portions of the Pacific Northwest, to the detriment of more rural areas. It was finally decided to abandon the zipcode-based scoring in favor of less disaggregated county-level scoring that did not consider prevailing wind direction."

Others involved in the development of the EBI agreed that it became necessary to develop such a tool because of growing constraints on the program. Alex Barbarika, an agricultural economist at FSA, explained, "We were getting near the cap of the program back then and we needed a way to pick the best offers—the most cost-effective offers. Originally, we could accept just about everybody who made an acceptable bid. Before 1990, we were trying to fulfill a minimum bid. [Now] we were trying to decide who we could accept and who we wouldn't." Barbarika noted that the EBI initially was not as important in affecting decisions as anticipated because, shortly after its development, crop prices increased and there was not as much demand to place lands in the CRP.

Ralph Stephenson (currently director of USDA's Conservation and Environmental Programs Division) and Mike Linsenbigler, also of USDA, agree that the 1990 Farm Bill's indication that the CRP could focus on erosion, wildlife habitat, and water quality helped transform the program from supply reduction to environmental protection. The rise in agricultural prices also decreased the need for the government to attempt supply reduction. Stephenson and Linsenbigler indicated that FSA began to think about how it could develop a measure to "find bargain lands" that would increase the environmental protection through the CRP. High-level officials in FSA went to the deputy undersecretary of agriculture with the idea of developing an EBI. Once the secretary of agriculture signed off on the concept, the undersecretary convened a meeting with all major agencies that might have an interest in the standard and created an interagency committee to work through the problems in defining the EBI.

The discussions about the EBI involved economists, forest service experts, and staff with other diverse backgrounds. The committee decided to weight the three factors mentioned by Congress more heavily than other goals, such as air quality. Key concerns revolved around what type of data was available, so that parcels could be consistently evaluated. Some analysts pushed to emphasize endangered species, for example, but data on their habitats were not easily assembled. In the first years of the EBI, soil, water, wind, and other data records were kept only as hard copies and all calculations were done by hand. Those designing the EBI needed a formula that could be calculated manually for roughly 275,000 offers covering millions of acres of land—in six weeks. Linsenbigler estimated that it took between two and three million calculations for the government to evaluate the lands submitted in the 10th CRP signup.

The need for tractability was apparent to some involved in the development of the EBI. Paul Harte, an agricultural economist, explained, "My role regarding the EBI was to oversee implementation of the use of the EBI in the process. I was a team member and I lobbied other team members to agree to changes in the EBI to help us ensure it was administratively feasible." Harte recalled that, when the interagency taskforce of economists and conservation officials got together, there were some suggestions that sounded good in theory, but proved difficult to implement. Eventually, in the discussions, "we were able to limit some of the ideas to the realms that were most viable." Some proposals were refined as committee members realized that particular calculations would be subject to human error. Harte also pointed out to the group that, given the large size of some CRP fields, it made sense to collect primary, secondary, and tertiary soil samples to "give a better halo of soil samples than just a single soil type."

Early on in the agency discussions about the EBI, some officials favored using a cardinal approach that would assign dollar values to the environmental benefits generated when a section of cropland was enrolled in the CRP. Essentially, per Harte's account, some people wanted to say that if a field saved five ducks and five ducks were worth $90, then that was how the value was assigned to the rating of the field. It was "basically impossible to assign dollar values." Eventually, he said, "We got them to use an ordinal approach that uses a relative ranking of numbers and doesn't rank and assign dollar values." By using the relative ranking approach, they steered clear of "saying that so many pounds of soil erosion is worth so many dollars."

The initial use of the EBI let government officials see the relative attractiveness of bringing parcels into conservation, but the process left landowners with no indication of how judgments were really being made. Linsenbigler said, "The first EBI used grand, amorphous language," but landowners did not know what exact formula was used to score their lands. The EBI was truly a "black box," with government officials treating the formula and the process as a "proprietary and classified process." Harte used similar language, noting that early on that "the process within the county to score the EBI was manual, but we didn't collect much data in the first black-box days." Explaining what would happen with the applications collected at the county level, he noted that back in Washington they would match a conservation practice to a set of geographically-based computer databases to develop a rudimentary score. Osborn agreed that "the black box EBIs were formula based," with the formula kept out of public view.

1996: OPENING THE BLACK BOX

The passage of the 1996 Farm Bill offered the opportunity to revise the EBI. The new version of the EBI was a publicly-released scoring system, which allowed farmers to see exactly how the environmental benefits of their land were being evaluated in the signup process. As Harte related,

> I can tell you this, [the transition] from black box to white box, I initiated. The reasons we did this was that we wanted to use the EBI as a tool to drive producers to give us maximum environmental benefits at a minimum cost. We encouraged them [producers] to increase the environmental benefits they were willing to provide and the minimum amount of money they were willing to accept...
>
> We can convince producers to do better practices. We are pushing them to do more and we are encouraging them to accept less. The less you are willing to accept increases your points. It is an inverse relationship. The less cost the higher the points. That is a lever. You can't do that in a vague [black-box] statement. When they can see it happening and see the difference [they act differently].

He concluded that the transparent EBI system "gives the best offers overall...relative to cost." In earlier discussions of the EBI, there had been proposals to try and divide benefits by costs. The difficulty of quantifying the dollar value of benefits meant that eventually cost was simply incorporated as another ranking factor, with a low bid price treated as a desirable by assigning the parcel higher ranking points

Parks Shackelford, assistant deputy administrator and then associate administrator in FSA in the 1990s, was heavily involved in the 1996 EBI revisions. He noted that one impetus for the changes in the EBI in 1996, after the passage of the farm bill, was that during the upcoming 15th signup "so many acres were expiring that it was going to be one of the first times for large amounts of land to go back in [the CRP]." USDA "needed a better way to insure that we were enrolling land that provided environmental benefits for the cost." To build the new EBI, he continued, "We went through a process of bringing together the conservation and wildlife communities and all of the stakeholders we could think of." The meetings focused on ways to come up "with soil erosion factors and ways to determine water quality benefits, wildlife benefits, and in some cases air quality benefits and then cost."

Some approaches built on the previous black-box EBI. Soil erosion rates had been central in the earlier formulas, but the new 1996 EBI adopted changes in this area, too. More attention was paid to differences in the type of vegetation, for example, distinguishing between native and non-native grasses and specifying types of trees to plant. The push for specificity also made clear the likely impacts of the scoring system. Shackelford noted that, in the meetings about scoring, "there is always strong disagreement. When you come up with something conceptually you have to balance a lot of interpretations of what that means and how it should work." Noting how bureaucrats pushed to represent their constituents' interests, he said, "If you are from Iowa, you are concerned about what is good for Iowa. FSA and the NRCS are agencies that have offices in every state in the union." This meant that people would reason backward from how a measure might affect their geographic area and try to fashion the scoring that would benefit them. "You can look at any of the criteria [this way]. If you are from an area with wind, you want more emphasis on wind erosion."

Shackelford described the 1996 revision as involving a wide range of participants from multiple levels of government.

> We would do regular briefings on how we were determining it [EBI]. The secretary's chief of staff was involved. We had a working group between the FSA, NSCA, and Fish and Wildlife, and the Forest Service...We brought in people from the field. We brought in people from different regions of the country to develop the EBI. Once that was determined, they [i.e., the people in the state offices] had some flexibility to determine state priority areas. They could determine areas of the state they wanted to focus on...which would give them extra points.

He commented that extensive discussion went into the formulation of the rankings. "There were different levels of briefing papers [we would use] to try and make a decision on different factors. You would have a decision memo that would have pros and cons on the issue."

The secretary's chief of staff would at times sit in on these meetings. Although Secretary of Agriculture Dan Glickman did not become involved in the details of the revision of the EBI, he did make the final determination of the EBI cutoff that would be used to determine which lands would qualify for inclusion in the CRP. For the 15th signup, the first after the open EBI was implemented, this enrollment was controversial. Shackelford noted that "the FSA wanted to enroll more acres and the NRCS wanted less [*sic*] acres." When the secretary ended up siding with the NRCS and choosing a higher EBI so that fewer acres would be eligible for the CRP, a "political firestorm" erupted.

The agency adopted a transparent EBI in 1996 because farmers had been so frustrated with the black box nature of the earlier EBI formula. Per Barbarika, "there was a lot of anger over the 1990 EBI because offices looked dumb." He noted that Congress was not heavily involved in the initial push for an open EBI, and that the original creation of the EBI was also not something that Congress directed. He recalled going to all the 1990 congressional hearings and markup sessions and the words "environmental benefits index" did not come up once.

The process of changing the EBI did not go through the rulemaking process of publishing the draft scoring methods in the *Federal Register*. Instead, the publication of the EBI method in agency documents and throughout the county offices allowed farmers, interest groups, regulators, and members of Congress see how decisions for CRP were made. With this information, farmers could now choose different types of land, alter conservation plans, and calculate rental prices in order to get into the CRP. More information and feedback began to flow back to those drafting new iterations of the EBI that accompanied the annual signups for the CRP.

Some of the changes in the EBI over time reflected data availability and access to technology. Ralph Heimlich described the centralized and dispersed nature of the EBI.

> Generally speaking, EBI is a national framework implemented with parameters from the regional/local level. For example, a structure for points for wildlife

> cover was mandated at the national level, but what covers met the various point levels was left up to the State Technical Committees to decide. This has to be the case because what constitutes the "best" cover for wildlife varies from area to area depending on climate, soils, wildlife species, and a host of other factors. Another example is calculation of soil erosion. The equations for sheet and rill and wind erosion are uniform nationally, but depend on locally evaluated parameters unique to the climate, soil and slopes of particular areas.

Emphasizing the technological constraints on trying to implement an open scoring process that every office across the country could easily participate in, Heimlich noted,

> You have to understand that, as late as 1996, FSA offices in the field were barely computerized. FSA first had to develop a computer-based signup process, which was difficult enough in itself. However, by the early 2000s, it became obvious that there was a high error rate in information associated with each bid, so information in tables was the first to be made subject to automated lookups, followed by information on maps (priority areas, distances to streams, etc.)... [They were] the first to be brought onto computer rather than being looked up on paper tables or maps. The next generation of EBI (maybe in the late 2000s) will start to look at information beyond the parcel itself, particularly landscape information about what is around the parcel. That will require implementation of web-based GIS tools. You may think this kind of tool is easy to develop, but you have to understand the large number of FSA offices—2,300 or so—and the low level of technology generally available for these tasks, coupled with the huge numbers of bids—100,000 or more in a signup—submitted within just a few weeks.

Barbarika noted that a portion of the later changes in the 2002 iteration of the EBI came from the need to simplify and automate the process. Categories not easily automated at the time were sometimes dropped. He pointed to one example, where they removed the points based on distance to water and wetlands because it was the source of much human error. Some people measured distances from the center of a lake and others started from the closest shore.

CHANGES IN THE EBI

The revelation of how the scores were developed in the EBI also opened up the discussion to many agencies, groups, and individuals and their suggestions about how lands should be scored. Most people involved in making changes to the EBI saw this input, then as now, as supplying helpful information that might not otherwise be easily obtained by regulators. Some even stress that direction from politicians, although rarely offered, is helpful because weighting specific factors in the index is inherently subjective.

Shackelford was high enough up in the agency (then associate administrator at FSA) to appreciate the need to look outside for information and feedback. He declared that, ultimately, "we answer to Congress," and that "there were at least two or three [congressional] oversight meetings on CRP," which meant that they also got "letters and calls from people on their staff." He observed, too, that stakeholders generally have an influence over the development of the EBI through their provision of information. Thinking about stakeholders, he said, "Yes, they had a significant influence. They had concerns. If you are an intelligent government employee, you don't know everything. You try and talk to individuals who know more about something than you do." He included conservation, agricultural, hunting, and wildlife organizations in the set of groups that provided feedback on the EBI over time.

Harte saw that conflict over priorities often generated discussion and analysis that improved the EBI. Describing the evolution of the measure, he reported, "There was definitely interagency tension...There has always been squabbles for priority." Divisions within areas covered by the EBI were evident: "There is a sort of an overlapping realm in commercial forestry and conservation forestry and they come from it from both perspectives." In wildlife debates, there were disagreements between those favoring hunting versus those supporting conservation. Positions could be strongly articulated, since "environmental folks really care about what they do." EPA tended to focus on protecting areas from point source pollution, while others lobbied for attention to soil productivity. Harte welcomed the input from interest groups, noting, for example, "that is their job, right?—to push for ducks." He pointed out that there can be a bias in what views are represented by organized groups, that "there isn't a single water quality group. There are many who have that set as a goal, but they are into more broad resource conservation."

LOCAL FACTORS IN THE EBI

Overall the operation of the EBI allows a national policy to be modified with local input and data. The determination of what the best cover for wildlife was, a factor in the EBI, was left to decisionmakers at the state level. While EBI equations spelled out how to incorporate soil erosion into ratings, information on local climate and soil was used to calculate parameters used in the equations. As Ralph Heimlich explained, this meant that "geographical considerations were very important, but not explicitly hardwired into the EBI."

The geographic incidence of the CRP decisions continually generated scrutiny of the EBI. Barbarika commented that this political pressure "is always in there. It isn't going to be too obvious...some people don't like CRP, but the next county or state over may think it is the greatest thing. We are always sensitive to shifting things around regionally. I don't know if it has had anything to do with getting anyone elected or not." As an example of geographical comparisons generating discord, he brought up the continual friction between North Dakota and South Dakota. North Dakota has a much higher enrollment rate than South Dakota, which generates political disputes, even though he credits the difference in CRP signups to the fact that the soil in North Dakota is more erodible.

Barbarika also noted that state offices did not have a role in explicitly formulating the EBI, but their comments on its implementation provided feedback helpful in revisions. The state offices do influence which lands are chosen for the CRP because they can designate priority, water quality, and wildlife areas. CRP applicants get additional EBI points if their proposed parcels fall into these areas. Local offices can influence the mix of environmental benefits arising from the CRP, as well, since they recommend what mixtures of seed cover should be planted where, for example. Some states did try to game the system, with the problems of accurately designating wildlife habitats being a case in point. States tried to designate their entire state as an endangered species zone just because the bald eagle flew over the area.

Government officials and stakeholders involved in the EBI see it as both effective and imperfect. As one agency economist put it, the EBI shows that "it is okay for the government to try and save money when they are buying environmental goods." He reflected that in an ideal world one might try to examine whether environmental effects produced closer to populations might be valued more highly, but the difficulties in translating these values into dollars prevented that approach from being adopted. Given the imprecision that would be involved in dollar-denominated benefits, he wondered, "Do people value wildlife or air quality? Who knows?" Imprecision also arose in the many debates about the number of points to award to a particular type of environmental benefit. Since the 1996 Farm Bill singled out three dimensions in particular—soil erosion, water quality, and wildlife habitat—these factors ended up being equally weighted in the 1996 EBI.

HOW TO CALCULATE THE EBI

The 15th CRP signup (1997) was the first opportunity after the passage of the 1996 Farm Bill for farmers to bid to place their land in conservation contracts and take advantage of the transparent EBI formula. To understand the wide range of factors embodied in the EBI, consider the categories of calculations involved. The 1997 version of the EBI contained seven components plus additional subfactors.[9] Excluding the cost component, the highest EBI ranking score would be 400.

First Factor: Wildlife Habitat Benefits

The first factor totals 100 points. This score is by far the most complex, taking multiple subfactors (numbers 1–6, explained below) into account. The formula is:

Wildlife Habitat Benefits = (cover subfactor/50) x
(cover subfactor + subfactor 2 + 3 + 4 + 5 + 6) .

(Cover) Subfactor 1. Cover and practices beneficial to wildlife (50 points). The vegetative cover planted on CRP lands largely dictates the level of habitat benefits. Fifty points are mostly given for native grasses and legumes, mixed hardwood trees and pines, or hydrology restoration of wetlands. This total is determined by state FSA officials.

Subfactor 2. Federal or state threatened, endangered, or candidate species (15 points). Lands that are within range of an endangered, threatened, or candidate species may receive up to 15 points.

Subfactor 3. Wetlands priority (10 points). This subfactor takes into account the proximity of wetlands. Semi-flooded or permanently flooded wetlands within one mile of a site score 5 points. Wetlands, within or adjacent to a bid area, equal to more than 5% of the contract area score 10 points.

Subfactor 4. Proximity to other protected areas (10 points). Protected wildlife habitat near proposed CRP lands is the target of this subfactor. Protected areas within 1 mile of proposed lands score 5 points and adjacent land scores 10 points.

Subfactor 5. Contract size (5 points). Larger tracts of land are better for animals, in contrast to small, broken-up, or scattered pieces of land. Tracts of land that are twice (or more) the average size of state/area contracts score 5 points and land parcels that are equal or larger than the state/area average, but less than twice the average, score 2 points.

Subfactor 6. Upland-to-wetland ratio (10 points). This subfactor is based on the ratio of uplands to the acreage of restored wetlands in the offered contract area.

Factor 2: Water Quality Benefits From Reduced Erosion, Runoff, and Leaching

The second factor has the same weight as wildlife habitat and also is worth 100 points. It is less complex, with only four subfactors that are simply added together to calculate the final score:

Water Quality Benefits = (Subfactor 1 + 2 + 3 + 4).

Subfactor 1. Location where crop production contributes to groundwater or surface water quality impairment (30 points). All 30 points are awarded if the area 1) is identified by the state and local water quality plans; 2) is in a state-identified wellhead and groundwater recharge area; 3) is covered by the *Coastal Zone Management Act,*[10] as amended, and its coastal nonpoint pollution control programs; or 4) is identified in plans developed in accordance with the Clean Water Act.[11]

Subfactor 2. Ground water quality benefits (20 points). This subfactor awards points according to the "leach index," which characterizes the potential for downward movement of nutrients and pesticides in specific soils.

Subfactor 3. Surface water quality benefits (40 points). This evaluates the potential amount of sediment that is delivered to water sources, along with populations within the watershed that would benefit from improved water quality.

Subfactor 4. Water quality improvements associated with wetlands (10 points). Bids receive 10 points if at least 10% of the land meets cropped-wetland eligibility criteria.

Factor 3: On-farm Benefits of Reduced Erosion

Erosion is scored through the Erodibility Index, as it has been since the inception of the CRP. EI is the sole criterion in determining how many points are received for erosion. For example, an EI score of 8 scores 25 points. Each additional point (from 9 to 19 points) gained in the EI scores 5 additional points, up to a maximum of 100 points. For example, an EI of 9 receives 30 points; an EI of 19 scores 80 points; an EI of 20 scores 90 points; and an EI of 21 and higher scores 100 points.

Factor 4: Likely Long-term Benefits Beyond CRP Contract Formula

The fourth factor has 50 points. It is a mix of 1) retention, or the likelihood that the practice established will persist and be maintained beyond the expiration of the CRP contract; and 2) the potential to reduce and sequester greenhouse gas emissions. Planting hardwood trees, for example, receives all 50 points because it has been shown that 80–92% of tree-planted land has remained in trees 10 years after the CRP contract expired. Re-established wetlands receive 25 points, and land planted in shrubs only gets 5 points.

Factor 5: Air Quality Benefits From Reduced Wind Erosion

The fifth factor is worth 25 points. The score is based in part on the inverse distance-weighted population potentially affected by wind-borne dust.

Factor 6: Benefits of Enrollment in Conservation Priority Areas

The sixth factor totals 25 points. Conservation priority areas (CPAs) are lands with high intrinsic value of resources. The full 25 points is rewarded if bid land is located within either a national- or state-designated CPA.

Factor 7: Cost

Cost is weighted after all bids are received and its weight may fluctuate, based on bid data. Cost is calculated from two subfactors, which are simply added together in the equation:

$$\text{Cost} = (\text{Subfactor } 1 + 2)\ .$$

Subfactor 1. Cost factor (*x* points). Suppose x were eventually declared to be 100 points. Then, the example provided by the agency in its EBI explanation indicates that the following formula would be used to calculate the cost factor:

$$\text{Cost Factor} = 100 - 100^{*}(\text{bid amount}/160)\ ,$$

where x = the number of points allocated for the cost factor.

Subfactor 2. Cost share (10 points). If cost-share assistance is *not* requested, the bid receives the full 10 points. This gives a small advantage to lands previously enrolled in the CRP, which do not require new plantings. If cost-share assistance is requested, the bid receives 0 points under this subfactor.

Overall, the EBI equation for the 15th signup was:

EBI = Wildlife Habitat Benefits + Water Quality Benefits from Reduced Erosion, Runoff, and Leaching + On-farm Benefits of Reduced Erosion + Likely Long-term Benefits beyond the CRP Contract Period + Air Quality Benefits from Reduced Wind Erosion + Benefits of Enrollment in CPAs + Cost .

Although this new calculation of the EBI was complex, for the first time, farmers could see how their lands scored on the agency's evaluation standard for environmental benefits. After all bids were received in the 15th signup and EBI scores were calculated, those lands with a minimum EBI score of 259 or higher were accepted.[12] Of the 23.3 million acres offered by producers, USDA accepted 16.1 million acres for enrollment in the CRP. The average EBI score for land enrolled was 307, 46% higher than the 210-point EBI average of the previously enrolled CRP acres. The theoretical maximum of the EBI in the 15th signup was 600 points, based on the sum of the six environmental factors and a possible score of 200 for cost.[13]

CHANGES TO THE EBI FACTORS AND SUBFACTORS BETWEEN THE 15TH AND 16TH SIGNUPS (1997)

Regulatory agencies will often publish decision rules and matrices similar to the EBI in the *Federal Register.* However, the officials running the CRP chose instead to print the EBI in agency circulars for each signup, rather than going through the extensive notice-and-comment rulemaking for each version of the EBI. This preserved agency discretion and allowed the EBI to be changed with each iteration and signup. After the 15th signup in 1997, for example, an interagency task force was put together to make alterations in the EBI for the next round of signups. The task force described and implemented the following changes for the 16th signup.[14]

Changes to the Wildlife Cover Sub-factor

EBI now began to award points for up to five different species of cover instead of points for mixed stands. This gave a better definition and helped differentiate landowners who were willing to adopt covers for wildlife habitats. This subfactor was also changed from giving points for predominance of cover (over half of the enrolled acres) to points based on the state of the cover. The minimum acreage of cover for scoring purposes became 51% for existing covers, 70% for a mixture of existing and new covers, 90% for new covers, and 100% for trees. This new system better recognized the value of existing covers and was most likely to save on cost-share payments for establishing new covers.

Changes to the Enduring Benefits Factor

There were four changes made to this factor.

1. **Points awarded for restoration of rare and declining habitat.** The preservation of these habitats is strongly linked to the survival of threatened or endangered species.
2. **Points awarded for cultural resource areas, such as historic sites and certain tribal lands.** This change is more consistent with other laws recognizing historic and cultural resources.
3. **Points awarded for planting shrubs.** Shrubs are now recognized as viable habitat for some wildlife, despite chronic underuse in CRP lands.

4. **Points can be awarded for non-CRP obligations to maintain CRP practices after CRP contracts expire.** This change recognizes efforts of state governments and organizations, such as The Nature Conservancy.

Changes to the Air Quality Factor

This factor shifted to a weight of 35 points instead of 25. It is also now subdivided into three subfactors. These factors were expected to benefit western states, such as Washington, Texas, and Colorado.

Subfactor 1. Wind erosion impacts (25 points). The change abandons analysis by zipcode in favor of county-based wind erosion and distance-weighted population factors. This revision gives more consideration to rural areas.

Subfactor 2. Wind erosion soils (5 points). This awards points to soils with a high percentage of fine material that is more likely to become suspended in the air.

Subfactor 3. Air quality zones (5 points). This subfactor evaluates areas where agriculture impacts air quality or areas that are located within 50 miles of a Class 1 air-quality area, such as national parks with high-quality air standards.

Changes to the Cost Factor

Another cost subfactor was added to provide points for offers of less than the maximum rental rate for soils in the offer. It awarded a point for every dollar below the maximum rental rate, up to 15 points. This change may benefit producers in areas of higher cost land, such as the Corn Belt and the Great Lakes States.

The ease with which USDA could change the EBI between signups shows how the agency could fine-tune the metric as it shifted priorities and saw how the EBI governed which lands were enrolled in the CRP.

Producers interested in proposing lands for conservation rental in general CRP signups had to submit information for the EBI scoring process. There are, however, other ways for lands to be enrolled in the CRP without going through the EBI bidding process. If the proposed land is seen as environmentally desirable and the producers agree to a particular type of conservation practice, the land may be automatically accepted under the CRP continuous signup policy. This option, however, only accounts for a relatively small percentage of lands that make their way into the CRP. In the October 2008 inventory of the 33.6 million acres in the CRP, 29.5 million acres had come in through the general signup process versus 4 million through the continuous signup programs.[15]

REGULATORY PATTERNS

Many regulations emerge from the familiar process of rulemaking. A notice of proposed rulemaking is announced in the *Federal Register*. Next follows the publication in the *Federal Register* of the proposed rule, with time for comments and studies from affected parties, who submit their analyses to the agency. The final regulation published in the *Federal Register* explains how the agency has responded to comments made during the rulemaking process and provides the text of the regulations that then bind the actions of the private and public decisionmakers subject to the rule.

The development of the Environmental Benefits Index shows how different the operation of the Conservation Reserve Program is from most regulatory programs. The EBI has not been published in the *Federal Register* for comments. The initial version was developed and used without the scoring formulas made public. The 1996 revision was circulated in agency handbooks and documents, so that farmers could see how their lands would be scored. The process of revising the EBI included input from stakeholders, and the ability of the agency to change the EBI from signup to signup allowed it to respond to changing priorities in environmental protection. The EBI development process appears to work well, in part because it is flexible and less cumbersome than the normal rulemaking process.

But the question remains: why can the USDA develop the EBI without using the normal regulatory proceedings used for federal rules? Part of the answer lies in the fact that farmers are not compelled to set aside their croplands for conservation; it is voluntary. If they choose to participate in the program, the EBI defines how the government evaluates their parcels. For many of the officials involved, this means that the CRP is not a regulatory policy.

Both Stephenson and Linsenbigler at USDA concur that the CRP is not a regulatory program, since regulatory programs force particular types of behaviors. As Stephenson said, "In our way of seeing the world, regulatory means programs that force action...CRP is voluntary in nature. EPA achieves results because it takes you to court and forces you to. We achieve the results we have through voluntary cooperation—through collaboration. It is an entirely different method of achieving results. We think we are very successful at it. [The U.S. Department of] Interior has credited the CRP with the recovery of a number of species."

Stephenson noted that the flexibility of the program means that factors can be inserted as priorities change. When global warming appeared on the environmental agenda, USDA created a factor for carbon sequestration in the EBI. He credited this flexibility with making the CRP the largest carbon sequestration in the country.

After the passage of the 1996 Farm Bill, the CRP underwent the rulemaking process, which generated comments on the types of environmental benefits involved in deciding which parcels to admit into the program.[16] However, the EBI was not published for direct comment in the *Federal Register* because it was not subject to *Federal Register* feedback for each iteration of the CRP signups. Shackelford observed that by not putting the EBI into a final rule the agency could "maintain the flexibility to make changes as needed." Harte explained how easy it is to modify the EBI, that "when we change the rule, we don't really go through that process...we don't even pay attention to the rule. The criteria is [*sic*] done through administrative discretion and not through rulemaking. We just have to change the handbook."

Harte further noted that this freedom from the rulemaking process does not mean that the agency simply acts on its own. The statutory requirement to take water quality, wildlife, and soil erosion into account when accepting lands into the CRP resulted in these factors being heavily and equally weighted in the CRP. He also said, "Just by administering a program that is so large, we get a lot of comments and we are generally receptive to those." He credited review group meetings, county offices, and other government agencies as sources of feedback. Stressing the value of feedback in adjusting the scoring system, he reflected, "Most refinements have been through employee suggestions or comments we have received through the implementation of the EBI."

CHAPTER 3

INTERPRETING THE CONSERVATION RESERVE PROGRAM IN THE FIELD(S)

Many regulatory programs deal with a universe of several thousand manufacturing plants that need inspection or hundreds of companies in a particular industry that must be monitored. The scale of the Conservation Reserve Program, however, is an order of magnitude larger than many regulatory programs. In 2008, there were more than 33 million acres enrolled in the CRP. USDA had to track nearly 740,000 rental contacts for more than 400,000 farms—contracts that generated $1.7 billion in rental payments in 2008.[1] Ideally, regulatory compliance in this program means that landowners understand the terms of the CRP contract, correctly interpret the questions involved in the Environmental Benefits Index, and offer bids that result in lands with relatively high environmental benefits entering the CRP.

How could things go awry in the CRP? Questions in the EBI that appear precise in Washington may be opaque in the fields. Landowners may have incentives to exaggerate what benefits their land will contribute or what their conservation practices will be. State agricultural employees, hoping to support what is a vital part of the state economy in some areas, may contribute to generous regulatory interpretations. All of this can give rise to inflated EBI scores. The complexity of calculations and imperfect data available in the initial signups, even with a transparent EBI, can give rise to honest errors in calculations. Evasion can also be a factor, where producers attempt to circumvent payment limitations by splitting up payments across convoluted ownership structures.

These factors, some inadvertent and some calculated, can produce false positives and false negatives. False positives in the CRP are lands that make it into the conservation program, but ultimately do not contribute many environmental benefits because the nature of the land or the conservation practices were misestimated or misrepresented. False negatives are parcels that miss the EBI cutoff and thus fail to be protected by conservation efforts: if their attributes had been correctly evaluated, agency officials in Washington would have wanted the land to be rented in the CRP.

USDA thus faces a tremendous regulatory quandary: how can it determine whether the CRP is operating well? If information were costless, this would not be a problem. USDA could instantly assess the nature of a field, its score on the EBI, and the likelihood that a parcel is correctly placed in or out of the CRP program. Yet, the transaction costs of investigating whether a 100-acre parcel is correctly scored on the EBI could be high, involving a trip to the field to collect soil information and assessments of conservation practices over time. Multiply this across the thousands of contracts struck during a general CRP signup period, and the problem becomes glaringly obvious if CRP officials spend too much time investigating CRP decisions.

Generally in regulatory programs, officials adopt two different styles of monitoring compliance in the field. Some opt for police patrols and, as the metaphor implies, these officials will send out inspectors to look at a sample of reality. In the CRP, this means looking at a sample of contracts and analyzing how the calculations were made and whether the EBI truly represents the actual state of a particular parcel. Such inspections are one way for USDA to discover potential problems with interpretations of the EBI. Officials in the field also have an incentive to pay some degree of attention to regulatory compliance as they help implement the CRP, if they know they may be audited. Producers are so numerous, relative to inspectors, that the odds of detecting violations are relatively low at the farm level. However, the actions of the auditors often spur creation of guidance memos and changes in EBI calculations that make EBI compliance easier. USDA police patrol monitoring frequently takes the form of regional CRP audits by its Office of the Inspector General.

The alternative form of monitoring involves someone "pulling a fire alarm," which triggers an investigation. USDA Office of the Inspector General maintains a hotline, which people can call to complain about the failure of others to comply with rules. A number of these calls deal with the CRP. This chapter shows what USDA is able to learn about the operation of the CRP in the field through these two types of regulatory monitoring, police patrol and fire alarm.

POLICE PATROL

Often at the start of a regulatory program, compliance is uncertain, in part because requirements are still being understood and in part because regulators themselves are new to the program. USDA's Office of the Inspector General in 1990 audited the initial regional implementation of the CRP through the Agricultural Stabilization and Conservation Service (ASCS), the forerunner of FSA. In California, the auditors focused on San Luis Obispo County because it had the highest level of participation in the state and looked specifically at the contracts of five landowners.

The ASCS San Luis Obispo County office made a number of errors in determining payments for four of the five producers or related parties sampled. The ASCS county committee approved CRP contract modifications that incorrectly allowed payments in excess of the original contract amounts; the county office accepted land into the CRP that did not meet the eligibility requirements; the county committee did not properly consider ASCS regulations on payment amount limitations to a producer or to the same family; and the county office neglected to post 1986 CRP payments in contract payment summaries. As a result, the payments made to these four landowners and/or related individuals or entities, totaling $620,043 for program years 1986 through 1989, came under question.[2]

Although nationwide audits in 1990 found that officials administering the CRP did not fully follow directives in the program, at times the violations were the result of strategic, willful actions by farmers. An audit of CRP contracts in two Texas counties and one Colorado county "disclosed that members of a family group misrepresented and/or concealed pertinent information about their farming operations to evade the $50,000-maximum payment limitation. One family member in Moore County, Texas, did not meet CRP eligibility requirements as he was not a bona fide owner or operator of the land prior to January 1, 1985. Actual and potential improper payments total about $1.5 million over the term of the CRP contracts (1987–1996)."[3]

As the CRP evolved, it focused on attracting more environmentally-beneficial land into the program, and audits looked more closely at the environmental quality of the land. In the days of the "black box" EBI, when the formula used was held closely within the CRP in Washington, the Inspector General's office also did not widely circulate results of its audits. The audits, obtained via a Freedom of Information Act request, came with notices stating, "This report is provided to program officials solely for their official use. Further distribution or release of this information is not authorized."[4] Audits in 1995 and 1996 revealed that there were multiple problems with how the agencies scored environmental benefits and implemented conservation practices in the field. The audit findings were often paired with recommendations for changes in agency behavior.

The review disclosed that the Natural Resources Conservation Service did not always prepare conservation plans that would reduce erosion rates to the soil-loss tolerance level as required and that some producers were not following the plans developed by the NRCS. It recommended that NCRS clarify the agency position that released acres must be planned to meet the soil-loss tolerance.[5]

The data used by the Farm Service Agency to evaluate bids submitted for the 13th signup was inaccurate in some cases. The investigators wrote pointedly about the problems they found.

> We believe that the data problems identified occurred because quality reviews were insufficient, procedures did not provide adequate guidance, and information available in the database was not fully utilized to verify eligibility and consistency. Also, the planned conservation practices were not always consistent with the types of environmental or conservation concerns identified... We recommend that FSA improve its quality reviews of CRP bid data prior to contract approval. We further recommend that FSA coordinate with the

> Natural Resources Conservation Service to develop a cohesive policy for conservation practice approval and consider development of an automated data-entry check to identify planned practices which appear inconsistent with the types of erosion factors reported.[6]

The difficulties in assembling EBI bids in 1995 were clear in other audits as well, such as the CRP review in Iowa County, WI. Investigators there estimated that 16,088 of the 46,543 acres in the county that were enrolled in the CRP were actually ineligible for the program. They found multiple errors in the enrollment process, such as "eligibility not being determined on a field-by-field basis; the use of inaccurate soil map units, cropping factors, and conservation practice factors; and missing documentation."[7]

The 15th signup for the CRP in 1997 was the first enrollment process that used a transparent EBI that producers could see in action. Lands were scored on seven quantitative factors, with a maximum score of 600 points possible for a parcel. There were more than 252,000 offers made, covering 23.3 million acres submitted during the signup process. Officials ultimately chose 259 points as the minimum EBI cutoff for a section of cropland to be accepted into the CRP, and 16.1 million acres made this cutoff.

As with any new approach to regulation, the initial implementation proved bumpy. Inspector General auditors found that

> the EBI computations were very complex and resulted in errors and inconsistencies in EBI points on 47 percent of the 91 worksheets we reviewed...We found that four States modified the point scores for various national ranking factors without approval, which contributed to the high error rate. Also, various software programs used by NRCS personnel to expedite the scoring process were not adequately tested and contained multiple errors that impacted the EBI score. Improper or inaccurate EBI scores affected whether a producer's offer was accepted or rejected.[8]

The auditors' conclusion noted that, as FSA recognized errors in implementation, the agency issued clarifying notices to agency offices and developed validation tests to improve the quality of the scoring process. The circulation of improved guidance documents to field offices is one way that regulators can increase the level of compliance over time as a rule becomes better understood.

Sometimes these police patrol actions were triggered by congressional attention. In 1997 a U.S. Representative from Washington State noted that in the 15th signup 21% of offered acres were accepted into the CRP in Washington, versus 83% in nearby Idaho, and 82% in Oregon. In response the Office of Inspector General investigated. The auditors found that differences in EBI scores in Washington versus Oregon were driven, in part, by real differences in the environments there. They also determined, however, that the guidance that Washington State NRCS officials provided to producers was conservative, for example, commenting that "producers were not encouraged to improve existing grass stands due to the low rainfall amounts in some areas." The auditors found that once the agency made changes to the EBI, "including improvements in the scoring criteria applicable to wildlife habitat cover…and air quality benefits from reduced wind erosion," then the acceptance rates for CRP proposals in the 16th signup became much closer between Washington and Oregon.[9]

The information that the audits by the Office of the Inspector General provided about the 15th and 16th signups generated changes in the CRP. Overall, these audits found errors in the EBI scores in over 40% of those offers reviewed. In response, FSA and the Natural Resources Conservation Service issued a joint agency handbook for this program. Changes were inserted in the EBI to make calculations clearer and easier. Provision was also made for second party reviews of all CRP offers. Despite these improvements, when the auditors examined the 18th signup, they still found EBI score errors in more than 40% of the offers examined. The auditors said that in 2000 the errors rose for at least three reasons. "The process is complex: NRCS field personnel are responsible for establishing scores for a total of 21 different factors and subfactors in conjunction with each CRP offer. Instructions are sometimes vague and confusing. Scoring criteria are imprecise."[10]

The auditors also noted that FSA was working on developing software that used GIS technology and data layers to reduce the likelihood of errors and improve consistency across the country. Ultimately, this highly automated approach to EBI scoring did get implemented. (See the current description of the EBI scoring process at the end of this chapter.)

FIRE ALARM

Another way that administrators in Washington, DC, learn about implementation of the CRP in the field is through "fire alarm" monitoring. Under this model of regulatory oversight, officials rely in part on complaints from the field to spot potential implementation issues. The image of fire alarm monitoring is clear: if someone spots a problem, they pull the alarm and attract the attention of outside regulators. In the case of a fire, the incentive to pull the alarm comes because individuals and their friends (or others) are at risk. A person's office, home, or neighborhood might burn down. With a government program as large as the CRP, an interesting question arises of who has the incentive to pull the alarm and protest about potential abuses. To the taxpayer, a better-run CRP does not translate into material, personal benefits. So why might someone pull the alarm in the CRP, and what would the complaint be?

Information to study this question came through Freedom of Information Act requests, seeking material from USDA investigations into the operation of the CRP. Some of the documents that came back were heavily redacted agency files that described how phone calls to the USDA Inspector General hotline[12] triggered investigations. Some of these calls came from people with a personal interest in how the program operated, such as government officials involved in administering the CRP. Although one caller was anonymous, the language of the complaint indicated that the person probably was a county USDA employee. The hotline summary of the call described the complaint as concerning "the rental rates offered under this program as outlined in Notice CRP 559...This was proposed to be market-based rates where in reality they do not reflect current market rates and are not based on actual or market rental rate data. Complainant is a county USDA employee. The rents are too low or too high. It is a major waste of the taxpayer money."[11]

Another hotline call suggested that employee complaints may have stemmed from anger about workload and consistency of interpretation. As another caller put it, "I would just like to know why NRCS in Alabama and Georgia won't [*sic*] required to do the second part of viewers of the CRP program as the rest of us were directed to do by the National directive. I think if the rest of us have to do this they should too."[13]

Some people who pull the fire alarm object to how the rules are being enforced and how the program is being operated. As a summary of a hotline complaint noted, "an anonymous complainant alleges that farmers (names not provided) in north central Missouri and southern Iowa are defrauding the

Conservation Reserve Program (CRP). The complainant says that farmers are buying land and 'putting it into the CRP.' These farmers do not take care of the land and allow the trees to die. The object of the program is to conserve the land, but all these farmers are interested in is 'receiving a check from the government and have land to hunt on.'"[14]

At times the complaints are specific, such as an allegation that the Yazoo-Mississippi Delta Levee Board was "inappropriately receiving Conservation Reserve Program and subsidy payments on the same tracts of land." This type of allegation generates a memo from the Office of the Inspector General to FSA, requesting that they provide information on the specific complaint. In this case, the allegation turned out not to be substantiated.[15]

Some times the calls deal with allegations of fraud. In a Texas case, a person called to say that a company was submitting bills to the government for work on CRP lands that were inflated. When FSA investigated, they determined that the charge was substantiated, but the low dollar values involved and difficulty of collecting records to prove the charge meant that the case was ultimately closed without action.[16] Allegations of fraud can even be the subject of high-level exchanges. When a constituent in Texas was concerned that a family participating in the CRP in Hudspeth County was allegedly committing fraud, he or she contacted Senator Kay Bailey Hutchinson (Texas) to complain. This triggered a letter from the senator to the USDA assistant secretary for congressional relations asking for a response to the allegation. FSA determined that the CRP participant was indeed eligible to participate in the program and had not engaged in wrongdoing. This news in turn was relayed in a letter from the Inspector General of the USDA to Senator Hutchinson. This is an example of fire alarm monitoring also serving as constituency service, since the senator gained credit for following up on the constituent's request.[17]

SIGNING UP

For a farmer hoping to place his cropland into the CRP, the road may start in a county office of the Farm Service Administration. The current, transparent design of the EBI and the advances in data storage and technology mean that landowners can now quickly see how their land will score on the EBI and how changes in conservation practices or to the land parcel might alter the probability of land being accepted.

Consider what would happen if a landowner walked into the North Carolina State Farm Service Agency office in Raleigh, NC, with questions about how a section of cropland would be scored in the EBI during a general signup period for the CRP. A database of satellite images allows FSA to call up a picture of the farmland. The agency can then superimpose a series of databases on the satellite images of the land:

1. **CLU database:** This is essentially a tax-parcel database that contains all of the property lines between fields. It allows a landowner to select the specific dimensions of the fields to be entered into the EBI scoring system.
2. **CP 36 database:** This database shows areas that have been selected at the national level for the planting of long-leaf pine trees (Long-Leaf Pine National Conservation Priority Area). If land falls within these areas, it is automatically eligible for the CRP.
3. **Water quality areas:** This database shows designated water quality areas, offering opportunities to improve water quality, and brings more desirable land into the CRP.
4. **State priority areas:** This database delineates the areas, chosen at the state level by the state FSA committee, to be set aside for conservation purposes. Fields that fall within these boundaries receive extra points on the EBI score.
5. **Soil-types database:** This is a large database of soil types. It can identify multiple types of soils in a single field and give the soil rental rate (SRR) of each type of soil.

Once producers choose a specific field to score in the EBI system, they can develop a scenario about what would happen if the land were in the CRP. The scenario includes data on many dimensions of the field: total number of acres, all soil types in the field, the total amount of acres of each soil type, the SRR for each kind of soil, percentage of land within a state priority area, percentage of land within the national long-leaf pine area, percentage of land within a water-quality zone, average water erodibility index (EI), the average wind erodibility index, rainfall factor, and climate factor. The scenario also lets producers within the long-leaf pine area in

North Carolina know whether or not the soils in the selected field will support long-leaf pine trees. Finally, the scenario includes the maximum payment rate per acre for the field and the maximum annual contract payment for the producer.

At an office as technologically advanced as the Farm Service Agency in Raleigh, a farmer can obtain color printouts of the scenarios, which include satellite photos of the fields with each different soil type delineated by a different color. A farmer can run these scenarios in different configurations to see how different combinations of cropland might be scored.

Once a scenario has been produced by the Conservation Reserve Program's GIS, the data is fed into another program that produces the EBI score for the offer. This second program is called the Conservation On-line System (COLS). The field information is entered from the scenario, as well as other information about the parcel, such as cropping history. Producers may choose different alternatives in setting up their bids, including whether they want to forego technical assistance payments, what type of land cover or conservation practice they will employ, and the minimum rental rate they will accept. All of these affect the EBI score. The transparent nature of the ranking allows producers to see what their score would be with different land characteristics or with and without technical assistance. Farmers can make different decisions about which parts of the fields go into conservation and what practices will be followed, all in an attempt to raise the EBI score. When proposing to place the land in the CRP, landowners will not know ahead of time what FSA's minimum cutoff score will be to accept lands in a given signup. This gives landowners an incentive to strive for a high EBI score, rather than simply hit a known cutoff point. The FSA office does, however, provide prior cutoff scores from previous signups, so that farmers may get a better sense for the EBI score needed to have land accepted into the CRP.

Once all the information is entered and producers choose the conservation practices promised for the land and rental rate they would accept, COLS automatically formulates an EBI score for each parcel of land submitted. Note that the rental payments do not have to flow to one person; the program allows multiple owners to be designated and receive payments. During the signup period, farmers will often take home printed GIS scenarios and EBI scores to discuss with family and coworkers, with an eye toward maximizing their likely payouts from the CRP. FSA officials see this mechanism as encouraging producers to implement the most environmentally beneficial practices at the lowest possible cost to the government.

This transparent EBI scoring system allows farmers to see right away how changes on the ground might affect their score. For example, a randomly selected field in the North Carolina database was initially scored at 254. Once the farmer cut out a section of field that had less environmentally-valuable soil, the EBI score increased to 277. The landowner can see how changes in each conservation practice will affect the EBI score. Since the government prefers to pay lower rental rates for the conservation lands, the scoring system awards one bonus EBI point (up to 15 points) for every dollar under the maximum rental rates listed for the parcel that the producer will accept.

The COLS data algorithms are sophisticated and targeted. Consider another field, chosen at random from the North Carolina FSA office. This field initially was not considered eligible, except for its location within the Long-Leaf Pine National Conservation Priority Area. This meant that, in order for this land to enter the CRP, it would have to be planted in long-leaf pine. However, when the whole field was scored by the EBI, the program detected that one type of soil in one section of the field would not support long-leaf pine trees. Consequently, the program would not generate an EBI number until that part of the field was eliminated from the proposal.

By design, local offices are not meant to have much discretion in determining EBI scores. Information and input from local officials does enter the scoring system values. At the state level, a committee established by presidential appointment is given the power to determine state priority zones. In a state such as North Carolina, this means officials can take into account water quality and the location of poultry and hog farms. Priority zones, for example, can be designated around these types of farms to reduce their negative spillovers.

County review teams, elected by local farmers, help determine the rental price for each soil type in each county. Each soil type has a soil rental rate, which is determined by applying the price of the predominant crop that a farmer would plant in that soil and the average yield received. For example, if a particular soil type is mostly used for corn in a county and one acre of corn can yield $41, then that soil type's SRR would be $41. FSA officials in North Carolina point out that this system puts states like North Carolina at a disadvantage because tobacco yields the best price for producers with most soils, but it is not allowed in this formulation.

Local control of the SRR has been changing recently. After complaints that neighboring counties were offering drastically different prices for land parcels, the national Natural Resources Conservation Service office created a national database of SRRs, in an attempt to generate more consistency across the entire CRP. While the SRR system used to be a bottom-up process, it has been evolving into a top-down system and deciding rental rates through the use of the Land Value Survey, which is generated by the county commission. This survey often focuses more on how much an acre of land is worth, which may generate more consistent estimates of the SRR. The SRRs are now also compared to neighboring counties and neighboring states to ensure consistency.

FSA officials currently state that the generation of an EBI score for a section of cropland can take less than hour. Before the use of the GIS database and COLS, FSA officials would calculate their part of the scoring and then send the offers to the NRCS for soil rental rates and other information to be entered into the bids. Unfortunately, the CRP was not as high a priority for the NRCS, so FSA requests often languished in a large backlog. The integration of databases and incorporation of GIS technology now means that landowners can see right away how their decisions about parcel composition, conservation practices, and rental rates affect their EBI score.

CHAPTER 4

THE MECHANICS OF MONITORING: GAO, CONGRESS, AND THE *FEDERAL REGISTER*

Although the passage of a farm bill is often covered in the media as the end of a long political battle, the true impact of the bill depends on how it is next translated into rules and regulations and applied in the field. In the case of the Conservation Reserve Program, members of Congress and the public have at least three different ways that they can learn about how regulators are carrying out this conservation policy. Congressional oversight committees may ask the Government Accountability Office (GAO) to develop a report on the CRP. Second, the congressional committees may call their own hearings on the CRP, which generate testimony and statements, plus data submitted to the committee by interest-group leaders. These oversight hearings are also a forum to quiz USDA officials about the program.

The *Federal Register* is a third opportunity for information exchange. USDA publishes proposed or interim rules in it to solicit comments on CRP policy and, when the final version of the CRP rules is published, it records how the agency responded to comments from interested parties. Although the Environmental Benefits Index has not been subject to formal rulemaking through the *Federal Register*, comments submitted by stakeholders in the rulemaking process provide the agency with feedback on how the CRP (including the EBI) can be improved.

GAO REPORTS

The Senate Committee on Agriculture, Nutrition, and Forestry and the House Committee on Agriculture are home to some of the most interested, and expert, members of Congress in the realm of agricultural policy. These committees are the primary oversight bodies for the CRP. When these legislators want to investigate what is actually going on in the field—how the CRP is implemented—they do not usually spend the time and resources of their own staff. They delegate the investigation to GAO, which prepares extensive reports on the CRP, using surveys, audits, and stakeholder interviews. The reports often appear as omnibus farm legislation is being debated and the operation of the CRP is open to redesign. GAO will also prepare reports in years when attention has shifted from general farm legislation, but the oversight committees still focus on how well the programs are being run. Topics of these GAO reports on the CRP show how the emphasis of the program shifts over time and how problems with its implementation have been addressed.

In response to a request in 1989 from Senator Patrick Leahy, chair of the Senate agriculture committee, GAO prepared a report examining the initial costs and benefits of the CRP. GAO found that the Agricultural Stabilization and Conservation Service (ASCS) had enrolled more than 28 million acres in the CRP by the end of 1988, making significant progress toward the goal established by the Food Security Act of 1985 of removing 40–45 million acres of cropland from production. This resulted in an estimated decrease of 574 million tons of soil erosion per year. GAO determined that USDA managers had primarily enrolled large numbers of acres, rather than focusing on the types of land brought into the program. To plant more acres with trees, USDA accepted less-erodible land. The agency failed to target the land at greatest risk for losing soil, which meant it also missed opportunities to improve water quality. In addition, the bidding process was not competitive. Even though the text of the 1988 fiscal year appropriations required USDA to limit payments to the prevailing local rental rates, GAO found that "USDA's instructions to local county offices allowed CRP rental rates in many areas of the county to continue at 200% to 300% of local rental rate."[1]

The 1990 Farm Bill directed USDA to focus more attention on improving water quality as it selected which lands to incorporate. In 1992, GAO issued a report that described how the operation of the CRP had changed. A three-step process determined which croplands offered by producers would be accepted by the program. County agricultural and conservation committees reviewed whether the land was eligible. The bid rental rate offered by the producer was compared to a bid cap, which depended in part on local rental rates. Eligible bids that were below the cap were then scored on the newly established EBI. As GAO described the procedure,

> all bids are ranked according to their environmental benefits per federal dollar cost to enroll them. Bids are accepted in rank order until the predetermined acreage enrollment goal is achieved. The total cost for each bid is its rental rate plus the estimated government cost-share to establish a cover crop. The environmental benefits calculation and ranking for each bid are based on estimated improvements in the following seven areas: surface water quality, ground water quality, soil productivity, conservation compliance assistance, tree planting, assistance to designated state water quality impairment areas, and conservation priority areas.[2]

GAO pointed out that USDA had committed nearly $19 billion in rental payments (and cost shares for conservation expenditures) to enroll 36.5 million acres in the CRP. The title of the 1992 report, "Cost-Effectiveness Is Uncertain," captured the theme that CRP's precise environmental benefits remained uncertain. The EBI formula was not made public and its calculations were not meant to translate into direct estimates of benefits. GAO also pointed out that USDA still brought in lands with lower environmental benefits because of its other goals of reducing commodity supplies and supporting farm incomes.

As Congress turned to debate the 1995 Farm Bill, the legislators asked for GAO's input. In anticipation of the farm bill debate, the Senate

Committee on Agriculture, Nutrition, and Forestry in 1994 sent out 1,047 letters to "food and farm organizations, environmental and conservation groups, and individuals asking for their input into crafting the 1995 Farm Bill."[3] Committee Chair Richard Lugar (Indiana) then asked GAO to tabulate the 135 responses. Of the 188 changes suggested in the letters, 52 focused on the environment and conservation. The three main themes emphasized were the need to continue the CRP, the need to increase the emphasis on water quality in enrolled lands, and the importance of redirecting farm payments from reducing production to raising the levels of environmental benefits.

GAO produced another report on the CRP for the Senate committee prior to deliberation of the 1995 Farm Bill. This report acknowledged that there was no magic source for land quality data at that time: "no comprehensive data exist to precisely identify the amount of CRP land and other cropland that is environmentally sensitive and should be kept out of production." The study focused attention on the idea of using parts of fields as "buffer zones" between streams and rivers and cropland. A grass strip left between a water source and cropland, for example, would shield some of the chemical and sediment runoff. The report ended with three suggestions for reform of the CRP.[4] First, modifying the CRP to focus more on creating buffer zones, rather than on removing whole fields from production, would reduce federal costs because less land would be involved. Second, per-acre costs could be reduced if farmers could earn revenue from environmentally compatible uses of CRP land, such as producing hay, in exchange for a lower rental payment. Finally, the CRP could provide more lasting environmental benefits if, instead of 10-year contracts, it purchased easements that restrict the use of the land for a longer period, such as 30 years, or permanently.

In between omnibus farm bill years, GAO examined more technical issues relating to the implementation of the CRP. In a 1999 report to the House Agriculture Committee, GAO noted that USDA's Natural Resources and Conservation Service had stopped providing technical assistance in the CRP program for six weeks because of budget constraints. Providing detailed data on the cost of information, the report found that NRCS charged USDA's Commodity Credit Corporation $98 for each parcel of land evaluated for the CRP and $456 per parcel to create a conservation plan once a section of land was accepted for enrollment.[5] The NRCS costs of providing this technical assistance were $49.2 million in 1998.

When funds for these studies briefly ran out, NRCS halted this work until appropriations by Congress restored funding. Note that by 1999 there were three ways for land to make it into the CRP: the general signups driven by the EBI scoring; the continuous, noncompetitive signup of land, such as filter strips or riparian buffers, which could protect water sources; and the Conservation Reserve Enhancement Program (CREP), which targeted areas, such as the Chesapeake Bay.

The CRP was also part of the discussion about the extensive staffing of FSA offices. A 1998 GAO report noted that for fiscal year 1997 there were 17,269 employees in FSA's national, state, and county (spread across 2,440 counties) offices. Salary and expenses for this level of administration was $956 million, with part of the effort spent distributing $7.4 billion in payments to 1.6 million farmers. The report noted that farmers appreciated the high degree of personal service in their interactions with local agents. In describing how producers interacted with FSA to enroll in the CRP, the GAO report found that a farmer usually made three visits to the local office to comply with the simplified application process. As FSA described how the newly transparent EBI procedure worked,

> On a farmer's first visit to enroll in the CRP, the farmer reviews an FSA map and indicates the tracts of land he or she is interested in enrolling in the program. The FSA staff enters the tract identification information on a CRP worksheet, and the farmer certifies that this information is correct. If the land is determined to be eligible for CRP, the farmer returns to the FSA office to indicate the rental rate he or she will bid and signs a CRP contract, agreeing to the terms and conditions set forth in the appendix to the contract. FSA staff enter the bid amount on the CRP contract, which the farmer signs. FSA selects bids from across the country. The farmers whose bids are accepted return to the county office to review and sign a conservation plan prepared by the NRCS.[6]

At times, GAO conducts its own surveys to inform Congress about the implementation of USDA programs, including the CRP. In preparation for the debate in 2002 over farm legislation renewal, GAO sent out surveys to NRCS state conservationists and members of the state technical committees, who advise them on conservation programs. The nearly 1,000 respondents generally rated USDA's conservation programs as effective.[7] Yet, in terms of improving water quality or protecting native species, these conservation experts rated some efforts more successful than others. The Conservation Reserve Enhancement Program, first authorized in 1996, allows the government to target specific lands for conservation rental contracts. The CRP continuous signup policy, started in 1996, gives USDA the opportunity to enroll areas, such as riparian buffers and filter strips, without going through the general CRP scoring and bidding process. As of 2001, 1.6 million acres of the CRP's 34 million acres were registered via these targeted programs. The technical committee members, however, indicated that the targeted CRP lands were more likely to yield these environmental benefits than the acres enrolled in the general signups.

The CRP played a supporting role in broader GAO agriculture and environmental studies. In a study of climate change, GAO noted that the CRP saved an estimated four million metric tons of emissions annually because of the carbon sequestration generated by its conservation methods.[8] Although some agriculture payment programs limit the amount a person can receive per year (such as the CRP's payment limit of $50,000 per person per year), GAO found that, in general, "individuals may circumvent the farm payment limitations because of weaknesses in FSA's regulations."[9] In studying environmental data tracked by the federal government, in 2005 GAO highlighted the information maintained by FSA on the approximately 690,000 active CRP contracts.[10] This database, housed in USDA's national computing center in Kansas City, held detailed information about the conservation practices on the nearly 35 million acres enrolled in the CRP at the time.

OVERSIGHT HEARINGS

Congressional hearings on a regulatory program serve multiple functions for legislators. They can hold agency leaders accountable by directly questioning them about the program. The hearings can be forums for interest groups and constituents to share information about the impact and implementation of a program in the field. When such hearings are held in a congressional district, they become part of the particular legislator's constituent service because local farmers and association members can interact with USDA officials. In the months leading up to farm bill reauthorization, the hearings become an arena for airing differing perspectives on how the legislation should be shaped. At a time when agency officials are contemplating the final rule for the CRP, these sessions are a way for legislators to make sure that the agency knows their reactions to proposed rules and regulatory interpretations.

When Georgia Senator Wyche Fowler brought the Subcommittee on Conservation Forestry of the Senate Agriculture Committee to Macon, GA, in January 1988, the hearing allowed those affected by the CRP throughout Georgia to voice their reactions to the relatively new program. At the start of the hearings, Senator Fowler announced that USDA was already making regulatory changes suggested by legislators to focus the program more on the environmental benefits generated by water quality. He said,

> I am pleased to be able to announce today that the United States Department of Agriculture and OMB [Office of Management and Budget] have already instituted some improvements we requested in the CRP...They will expand the cropland eligible for the Conservation Reserve Program, if it is planted in trees. The erosion rate required to sign land up for the reserve will be reduced from the 3T category to the 2T category...Previously, in order to sign up a field, two-thirds of it had to be classified as highly erodible. Under the new regulations, only one-third needs to meet this classification, if the producer agrees to plant trees. Hence, this rule change not only increases CRP tree planting, but helps many farmers achieve conservation compliance as well...the rule changes will allow cropland to be enrolled in the reserve without meeting any highly erodible criteria if it is situated next to a stream or a lake and can substantially reduce runoff if planted in grass, shrubs, or trees.[11]

The 1988 hearing revealed that, although the CRP originally was seen as a way to reduce commodity supplies (thereby raising prices) and support farm income, three years into the program USDA was already stressing the environmental benefits beyond a simple reduction in soil erosion. Those in the farm services industry, however, often opposed expansion of the CRP acreage because land not planted with crops meant reduced demand for farming inputs. Dennis Dozier, vice president of Peavy Farm Service in Camilla, GA, put a human face on this argument. He argued, "I become very concerned when Congress proposes long-term solutions to a short-term problem. The current surplus of grain is a short-term problem. We should be careful when we attempt to reduce the surplus of grain with a 10-year conservation reserve."[12] At a time when legislation had been introduced to expand the number of acres allowed into the CRP, Dozier forcefully made the point that "if set-asides are carried to the extreme, then the very infrastructure of rural Georgia could be crippled."

At the same hearing, Wilbur Dellinger, Georgia state chair of Quail Unlimited, described how hunters and others valued the wildlife benefits of

CRP habitats. Describing the use, existence, and bequest values in language more eloquent than usually found in official studies, he pointed out, "Wildlife, which includes both game and nongame species, does not belong specifically to the landowner. Wildlife belongs to the people. In order for future generations to be able to enjoy wildlife, either through hunting or just observation, there simply has to be some wildlife present." Dellinger also noted that farmers need to be compensated and led to incorporate wildlife habitat plans in their CRP conservation plans, saying that "if you think the average landowner would include wildlife habitat in their planting program without it being mandatory, I bet you would leave the light on for Jimmy Hoffa."[13]

When a joint hearing of the House Agriculture Subcommittee on Environment, Credit, and Rural Development and the Senate Agriculture Subcommittee on Agricultural Research, Conservation, and Forestry convened in Aberdeen, SD, in September 1994, the focus was on the future of the CRP. At a time when some acres were set to come out of the CRP—attention was concentrated on the federal deficit and legislators were starting to plan the components of the next farm bill—the hearing participants generally argued for renewal of the CRP because of its many environmental benefits. U.S. Representative Collin Peterson (Minnesota) announced that the CRP was a priority for the sportsmen's caucus in Congress.

> We are going to try to make this a political issue in that we are going to try to get members of Congress and candidates for Congress to take a position in this election in favor of extending the CRP, so that we have as much political support as we can have when we get in the next Congress and in the fight on the farm bill. And we are trying to marshal support outside of the agricultural community for extension of the CRP, primarily with hunting groups and fishing groups and others that are very interested."[14]

Echoing this support was Dave Nomsen of Pheasants Forever, a wildlife group working in a coalition supporting the CRP that included the Delta Waterfowl Foundation, Quail Unlimited, Ducks Unlimited, the Wildlife Management Institute, the Wildlife Society, and the Izaak Walton League. He noted that his group believed that the "Conservation Reserve Program has been the most successful farmland conservation initiative in the nation's history" and should be renewed for 10 more years.

Secretary of Agriculture Mike Espy traveled to the South Dakota hearing to voice his continued support for the CRP in the upcoming debates over the 1995 Farm Bill. Many of the participants in the hearing offered studies or statistics to support their perspectives on the CRP, and Espy did too.

> We had…done a survey, 1993, of 17,000 CRP participants. That is about 5 percent of those who participate in the program. And we asked them, assuming that the authority would not be extended, what would you do with the property? And the survey indicated that 63 percent intended to return their acreage to crop production; 23 percent to put in grass for hay production or forage for grazing livestock; 4 percent, in trees for commercial wood products; 2 percent, in grass; 3 percent, in grass with no anticipated use; and 3 percent said they would sell it. Well, this kind of information is exactly the kind of information that the American taxpayer needs to be presented with.[15]

Espy argued that, if the CRP program expired and acreage came back into crop production, this would end up increasing deficiency payments and drive up farm program outlays.

The oversight hearing held in September 1996 by the House Agriculture Committee shows how legislators attempt to influence regulators as they write proposed and final rules. The hearings began with Chair Pat Roberts (a Republican from Kansas) voicing his displeasure with the (Democratic) Secretary of Agriculture Dan Glickman. "…I am not a happy camper this morning. I am irritated and I am frustrated. I must say I am surprised that Secretary Glickman, according to press reports, found that campaigning for a House candidate in Indiana, against a member of this committee, was more important than discussing the administration's conservation policy with this committee. It is even more puzzling to me that Undersecretary Lyons, who has the primary responsibility for these programs, is also unable to appear today."[16]

Frustration with the secretary of agriculture was also evident in the questions of Republican Representative Frank Lucas (Oklahoma), who prefaced a question to a USDA official with the statement that, "since the Secretary [of agriculture] is not here and has not apparently been able to return my phone calls, I find it necessary to ask you."[17]

Part of the frustration on the congressional side stemmed from the time it took USDA to develop the interim final rule on the long-term implementation of the CRP. Legislators noted that the agency had missed at least one mandated deadline for the announcement of the new CRP program. With USDA announcing an interim final rule for the CRP in early September and asking for public comments on it, members of Congress pointed out that signup rules might not be implemented until March, which might be too late in the spring for those who did not make it into the program to plant certain crops.

Representative Peterson objected that wildlife benefits in the CRP were not emphasized sufficiently in the proposed rule version and noted, "We are going to be taking a look at putting together a proposal that we can bring forward to the Department [USDA] during the rulemaking process to try and see if we can get that established back in the program." If things are not spelled out in detail in legislation, the legislators will often find fault with the way that regulators formulate policy. Talking about his frustration with how wildlife benefits were, in his view, slighted in the interim rule, Peterson said of the delegation process,

> Frankly, I think it ought to have been there when this rule came out, and that is what most of us thought was going to happen. As you are aware, Mr. Chairman, as we went through this process we have these gnomes in the back room that kept adding things to this bill that we never could figure out exactly where they came from, and we had to scurry around and try to, you know, beat them back at the last minute. You know, you wonder where some of this stuff comes from, but it looks like the gnomes were still in the back room while they were putting this together because some of the things that we were concerned about ended up in the rule.[18]

The members of Congress in the hearing recognized that one of the most important factors in determining which lands will be brought into the CRP, the EBI, was not specifically spelled out in the rulemaking. Parks Shackelford (at the time, the assistant deputy administrator for farm programs at USDA) described this as an advantage because it left the agency with discretion to change the EBI across signups without bearing the transaction costs of rule-making. The exchange between Representative Wes Cooley (Oregon) and Shackelford points out how members of Congress at times can view agency discretion as arbitrary:

> *Cooley:* ...In your rulemaking process, there does not seem to be any criteria for the EBI rule. Could you explain to me where that is going to be defined, and how it is going to come about so that everybody can really understand what it means?
>
> ...

Shackelford: The EBI has evolved over the years. We used it in the last signup. We have learned some things from that. We will use it again in this signup. It is probably not best to put down a fixed formula in a rule that we cannot come back and change if we learn we need to do some other things to it. We make it very clear to producers when a producer goes in to sign up, that producer can tell. We can show him the score for his land, explain how it was done for that particular signup, so that is very clear and very open, and how it is ranked for environmental factors.

Cooley: So the producer has to come in and sit down and talk to you before he can find out exactly what the EBI means to him?

Shackelford: That is correct.

Cooley: So it is really kind of arbitrary, depending on a lot of criteria?

Shackelford: No, sir, it is not arbitrary. It is uniform in the way it is applied to all producers, but you have to look at that particular piece of ground before you can develop an EBI for it, and each parcel of land offered for CRP is different.

Cooley: Well, so it is kind of arbitrary just for the land itself.

Shackelford: No, sir, it is not arbitrary. Arbitrary means it changes for everyone. I mean it is specific for that land, but just as Mr. Combest brought up the issue earlier that he would like to have air quality considered, you know, that is one of the things that we need to look at taking into consideration. But it is science-based determination, but as each parcel of land is different, the environmental benefits of each piece of land are different.

Cooley: So it depends on who is doing the judgment is what you are saying then. Does the producer do the judgment, the local authorities, or does the USDA? Who makes the determination?

Shackelford: The USDA makes those judgments. USDA has experts who are trained to make those judgments.[19]

Shackelford explained that each eligible parcel was scored on the EBI, the secretary of agriculture would determine an EBI cutoff, and those lands that scored above the cutoff could be brought into the CRP through rental contract agreements. When members of Congress again asked if the EBI scoring system was part of the rulemaking process and subject to comments, he said, "No, this is not a part of the formal rule because, while we have worked on it, I cannot say that it is absolutely perfect and it cannot be improved. In each signup, we may be able to improve on it more."[20]

Throughout the hearings, members whose districts might be affected by the treatment of soil erosion, rental bid levels, and wildlife habitats in the signup process asked questions about these factors. As Representative

Tom Ewing openly stated, "we all approach this change in the CRP parochially, I guess. I get a lot of questions in Illinois as to how this is going to affect CRP ground in Illinois, and I know that like all states we have a wide variety of land, and it will affect it differently. Can you give me any overview? What do I tell my constituents or farmer from wherever in Illinois when they ask, "How is this going to apply to us as we change it?"[21]

The final rule on the CRP was released on February 19, 1997, with the next signup slated to begin in March. The House Agriculture Subcommittee on Forestry, Resource Conservation, and Research held oversight hearings a week later on February 26. While members of Congress praised the CRP, they faulted USDA for failing to issue the regulations for the CRP within 90 days of the farm bill's passage, as requested in the legislation. Representative Larry Combest (Texas) complained that "the complexity involved with this scoring process coupled with the lateness of the rule have led to much uncertainty and fear in the country."[22] Dallas Smith, acting undersecretary for Farm and Foreign Agriculture Services, noted that the policy development took time because of the extensive input USDA sought on the CRP. Forums on the CRP were held in each state, and the agency received over 3,400 comments on the proposed rule. In an attachment to his testimony, he provided the committee with a detailed description of how the EBI scoring system would work for the 15th signup (which was published in the hearing's formal report).

The ranking Democrat on the subcommittee, Representative Charles Stenholm (Texas), eventually reminded his fellow legislators, who appeared to second-guess USDA on the EBI about how they came to this point. In the hearing, he said,

> I would also like to remind my colleagues that this is the route we chose to take last year during debate on the farm bill. We simply reauthorized the program and left it up to the Department [USDA] to implement this program for years to come with a few paragraphs of report language as their guidance. I would like to point out that in simply reauthorizing the program, we delegated to the Department the daunting responsibility of seeking ways to implement the program in such a manner as to further increase support among all interest groups: environmental, commodity, wildlife, grain industry, etc. For those of us who support the program, it does no good to narrowly define our support. I encourage USDA to continue to reach out to all groups in implementing the program.[23]

As legislators worried that the forms and data requirements would overwhelm their farming constituents, Shackelford reassured them that FSA would be able to handle the volume of information and would provide the expertise. He noted, "All our FSA offices are connected to a central computer…So we get that data when we get the producer signed up. The majority of that stuff, as we've said repeatedly, is going to be filled out by USDA employees, and we'll get that information correct, and we upload that information into one central computer that will then process it and rank those bids."[24]

To the legislators who complained about the complexity of the scoring system, Shackelford had a clear response. He said, "Everybody wants it to be simple, but everybody wants particular concerns of their region. You want us to give points for wetlands. The chairman wants us to give points for air quality and wind erosion. We have to take all of those things into account, and we try to balance—assess the environmental benefits as much as we can and make it as simple as possible."[25]

The same House subcommittee reconvened in June 1997 to review how the CRP contracting had worked in the 15th signup (March 1997). Dallas Smith reported that USDA accepted more than 160,000 offers and brought in 16.1 million acres. Each parcel was scored on the EBI, which produced a ranking based on seven factors—soil erosion, water quality, wildlife habitat, air quality, location in a conservation priority area, long-run retention of the conservation benefits after the rental contract was over, and cost (e.g., the rental bid the producer would accept). He noted that the mean EBI score for lands accepted in the 15th signup was 46% higher than the acreage currently in the CRP. Yet, Congress members whose constituents complained about failing to make it into the CRP voiced questions about the process to USDA witnesses. Representative Lucas admitted he was representing a narrow view when he looked at a map showing fewer accepted CRP offers in his area.

> I appreciate that, representing the farmers and ranchers and all the citizens of the 6th District, I have an obligation to work on their behalf and defend their interest in how programs are implemented. You [USDA] gentleman represent agriculture and producers and the citizens of this entire nation. Just as I have a parochial interest to work on behalf of my constituents, you have an obligation to do things that not only fit the rules, but that are things that present a sense of fairness and appropriateness for the whole nation. Granted, I can't read the tea leaves of what went on over there when this decision was made, but

> whatever the process, the end result produced an awfully interesting map that I am having a real challenging time explaining back home.[26]

As producers and Congress members grew more familiar with the implementation of the EBI, fewer questions in oversight hearings focused on the scoring system. Legislators in hearings focused on topics, such as whether the secretary (of agriculture) should try to enroll more acres in the CRP (up to the statutory maximum) or leave room for high value lands offered in the next signup round. Discussion also centered on the other avenues for lands to make it into the CRP, such as the continuous signup and the Conservation Reserve Enhancement Program.[27] By the time of the Senate Agriculture, Nutrition, and Forestry hearings on agricultural conservation in 2006, the CRP had been in operation for such a long period that its benefits were now more easily estimated and enumerated. John Johnson, deputy administrator for the Farm Program at FSA, told the committee that

> total enrollment now stands at 36.7 million acres with annual rental payments to producers totaling $1.8 billion annually. These acres have reduced soil erosion by 450 million tons, reduced nitrogen, phosphorus, and sediment leaving the field by over 85 percent, and sequestered over 48 metric tons of carbon dioxide on an annual basis. CRP contributes to increased wildlife populations as well, including more than 2 million additional ducks annually in the Northern Prairie, recovered sage and sharp-tailed grouse populations in eastern Washington, increased ring-necked pheasant populations, as well as increased grassland bird populations.[28]

Knowing whether conservation policies are actually being followed on CRP acres can be a difficult task. FSA generally required county offices to run a compliance check on 10% of the local contracts each year. With contracts running 10 years, this meant that a parcel might be checked for compliance with conservation measures once during its rental cycle. When the agency sought comments through the *Federal Register* on how to handle contracts that were expiring, some of the 5,000-plus comments received dealt with compliance. When Senator Saxby Chambliss (Georgia) asked Johnson in a 2006 agriculture committee oversight hearing how the agency was dealing with compliance, Johnson reported that, for the upcoming extension or re-enrollment of 28 million acres, USDA would require a compliance check on every contract. In a "pay-as-you-go" measure, FSA instituted a $45 fee per contract, plus an additional $1 per-acre fee, with the funds going to hire temporary employees to help with the compliance workload. While most of the concern centered around noxious weeds growing on CRP land, Johnson admitted, "Anecdotally I've heard reports that we found at least one trailer park and one water tower on the CRP acre."[29]

RULEMAKING AND THE *FEDERAL REGISTER*

Once the Food Security Act of 1985 was passed, USDA's Agricultural Stabilization and Conservation Service issued an interim rule describing how the CRP would operate and inviting comments on this rule. The final rule, laying out the general roadmap for the CRP's operation, was issued in February 1987. Only 28 comments were received on the rule, with most (24) focused on how erosion would be considered when lands were proposed for inclusion in the program. The text of the legislation gave the secretary of agriculture the discretion to take into account the erosion and productivity of cropland acres that farmers proposed for inclusion in the CRP. The interim rule proposed a very specific definition for which lands would be eligible based on erosion, saying that "the highly erodible land must be in a field which has been determined to predominantly consist of land classified by the Soil and Conservation Service (SCS) as being Class II, III, IV, and V, with an average annual erosion rate of 2 times the soil loss tolerance ("T") or greater as announced by the Secretary [of agriculture], or land classified by the SCS as being Class VI, VII, or VIII."[30]

The comments offered different ways to take erosion into consideration, including the suggestion that potential for erosion be factored in (rather than having to wait for heavily damaging erosion to take place to qualify a cropland parcel). After reviewing the comments, USDA announced that eligibility would be based on both potential and actual erosion. The agency chose to use an Erodibility Index to estimate the potential for erosion and announced that land with an EI of 8 or higher would be identified as highly erodible. This is an example of how feedback from the regulated community can lead the agency to modify its approach through rulemaking.

One commenter asked that the Commodity Credit Corporation (CCC, the government-owned corporation within USDA and the official entity entering into contracts to pay producers in the CRP) publish a formula specifying the weights given to factors in deciding which lands to accept into the

CRP. USDA rejected this suggestion, noting that "it was also determined not to publish an established bid acceptance formula. Offers to participate in the CRP for 1986 were accepted on the basis of the lowest bids received, with a per-acre bid ceiling established by the Secretary [of agriculture] within a geographic area. It is necessary for the CCC to maintain flexibility in the offer acceptance process in order to ensure that the goals of the CRP are achieved by the most cost-effective means possible."[31]

When a commenter asked that the maximum acceptable rental rate be announced before the bids were submitted, USDA again argued that "it is important that the CCC maintain flexibility when establishing the offer process, so that improvements in the process can be implemented as needed. Information describing the offer process is available to interested persons at all state and county ASCS offices."[32]

The 1985 legislation did have some direct constraints on the operation of the CRP, such as the provision in section 1231(d) that said the secretary of agriculture could not place more than 25% of the cropland in a county into the CRP unless the secretary determined that it would not have an adverse impact on the local economy. The legislation specified that a maximum of 45 million acres could be enrolled. In general, however, the relatively few details spelled out in the legislation meant that USDA had discretion in how it defined the rules governing the CRP.

Part of the information flow generated in rulemaking stems from the executive order requiring agencies to analyze the benefits and costs of major rules. After the passage of the 1995 Farm Bill, USDA's FSA released a final rule in February 1997 that outlined the long-term policy for the CRP. To comply with the benefit-cost analysis requirement in Executive Order 12866, the agency reported its assessment in the *Federal Register* of the likely impact of the CRP if it enrolled 28 million or 36.4 million acres (the legal maximum then).

> Some of the environmental benefits that have been estimated and applied to the CRP enrollment scenarios include: soil productivity ($150 million annually for the 28-million-acre scenario and $195 million annually for the 36.4-million-acre scenario), improved water quality ($350 million and $455 million, respectively), and increased consumptive and nonconsumptive uses of wildlife ($1.5 billion and $2.0 billion, respectively). The sum of these three categories, which would only be a partial accounting of the environmental benefits, is $2.0 billion per year and $2.7 billion per year, for the 28-million-

> acre and 36.4-million-acre scenarios, respectively. Enrollment of 28 million acres and 36.4 million acres is expected to increase annual net farm income from production of feed grains, wheat, cotton, and soybeans, CRP payments, and production flexibility contract payments by about $5.8 billion and $7.6 billion, respectively, compared with the no-CRP-continuation scenario. The increased net farm income results from higher commodity prices, reduced production expenses, and higher CRP rental payments to participants... [this] results in total estimated annual benefits to society that exceed costs by $1.1 billion and $1.2 billion, respectively, for the 28.0-million-acre and 36.4-million-acre scenarios.[33]

As the CRP grew in size and experience, many more organizations participated in the rulemaking process. In announcing the CRP final rule in 1997, USDA noted that it had received 3,467 comments on the proposed CRP rule it had released in 1996. The agency received nearly 100 or more comments on many facets of the CRP, including the environmental benefits generated by the CRP, the use of the Erodibility Index and selection of 8 as the minimum score to define eligible lands, the setting of the local rental rates for different types of soils to be used in the bidding process, and recommendations for conservation priority areas in the program. For the controversial question of whether some types of hay planting and grazing should be allowed on CRP lands, the agency chose not to use its discretion. Rather, the CCC said it would go back to Congress for direction. As the final rule notice indicated, "in the view of the divergence of opinions expressed by respondents on how the provision should be implemented, the CCC will seek legislative amendments to modify the existing provisions relating to haying and grazing of CRP acreage and obtain specific authority for periodic managed haying and grazing. However, existing provisions of the 1985 Act generally prohibit the nonemergency haying or grazing of CRP acreage."[34]

While the EBI had already been devised and was set to be used in the 15th signup, the rule did not discuss how the EBI would be employed or how it was constructed. When the CCC issued the interim rule in 2003 implementing the CRP under the Farm Security and Rural Investment Act of 2002, the *Federal Register* text openly described how the agency had used the EBI to rank CRP parcels since the 10th signup in 1991. The agency praised the EBI along several dimensions, saying that "the goal of the EBI is to provide a relative rank order of submitted offers based on environmental factors and cost in a uniform and consistent manner for all offers. In addition, the EBI provides incentives to increase cost-effectiveness. Ultimately, the EBI is used to rank the anticipated environmental benefits from each CRP offer."[35]

This iteration of the CRP rules stressed how enrollment was being targeted to bring in environmentally sensitive lands and encourage use of native vegetation to provide better wildlife habitats. The interim rule description noted how the agency aimed to do this through the discretion it had in calculating the EBI, and how a change in the index resulted in changes in producer behavior.

> Because native grasses generally offer better habitat than introduced grasses, the FSA has revised the EBI selection criteria to give greater weight to the use of native seed and vegetation species. This provides incentives for producers to offer and use native species in their CRP contracts. The effect of the greater EBI weight given for using species can be seen by examining the change in the proportion of CRP-enrolled acres planted in native grasses. In 1993, when there were nearly 30 million acres of grass in the CRP, only 28% of these acres were planted in native grasses. Since 1998, 67% of the grasses established under the new CRP contracts [were] native grasses.[36]

In the final rule published in May 2004, USDA noted that it had received 800 comments on the CRP. Many dealt with perennially controversial topics, such as land eligibility and the permitted uses of CRP land while under contract. In response to questions about protecting fragile lands and encouraging biodiversity, the CCC stressed that it was using its regulatory discretion to bring about changes in implementation by changing the EBI. The CCC stated in the *Federal Register* that "the 2002 Act made no changes in the EBI. However, in an effort to maximize environmental benefits and implement plantings consistent with local ecosystems, CCC has structured the EBI to give more weight to contract offers that devote acreage to native plantings."[37]

One commenter suggested that the reliance on the EBI—a formula published in agency guidance documents rather than hammered out through the notice-and-comment rulemaking process as part of *Federal Register* publication—was dangerous. The use of an informal rule, such as the EBI, could raise the chance that a disgruntled producer might challenge the system. The CCC appreciated the suggestion, but stressed that the agency preferred to keep the EBI out of a formal regulation.

> Another comment recommended the EBI be included in the regulation to protect the EBI against legal challenge. The agency understands the respondent's

> concerns that the acceptability of offers be more strictly determined by regulations. The weighing of factors in the EBI can change over time and over enrollments based on changing conditions, changing needs, and based on the nature of the land achieved in previous enrollments. Incorporating the EBI into the rules could harden the index in a way that would be harmful to the achievement of the goals of the program because of the time that would otherwise be needed to change the index.[38]

In essence, the relative silence of farm bill legislation on how USDA should choose which lands to enroll in the CRP meant that the department preferred to use clear but flexible rules to evaluate lands for inclusion. By keeping the decision metric out of the formal rulemaking process, USDA was able over time to alter the priority it gave to different types of environmental benefits without having to engage in extensive consultation with stakeholders about each change.

CHAPTER 5

THE ENVIRONMENTAL WORKING GROUP PULLS THE PIECES TOGETHER

If Congress gives out billions of dollars as part of nearly 740,000 conservation rental agreements with farmers, it needs to know who gets how much each year.[1] USDA has up-to-date information on who receives payments in the Conservation Reserve Program, as well as who gets the more than $11 billion a year in subsidy payments from other commodity programs.[2] USDA's Farm Service Agency or Commodity Credit Corporation does not, however, simply post this information on the internet. In part, the discretion of an executive branch agency lies in the power of hidden action and hidden information. If the agency made its payment data public, down to the level of individual checks, members of Congress might not like the distribution of funds. Winners and losers in congressional districts would be readily apparent. Winners and losers among interest groups, too, would be obvious, with the distribution of payments to particular producers visible from payment data. Members of Congress might not appreciate such scrutiny by disgruntled constituents and lobbyists. FSA officials would not welcome the public or legislative second-guessing of the pattern of their distributions. And individual farmers would not like web surfers or their next-door neighbors seeing how much they are getting from the government.

Imagine, though, what might happen if one could get into USDA data on individual payments to farm producers. It would be easy to see if a relatively small number of congressional districts had a high concentration of commodity subsidy payments or just how much large agricultural producers receive in commodity subsidy payments, or the relatively wide dispersion of the relatively smaller payments that go to farmers under the CRP. These patterns might help change the terms of congressional debate because the public might see that a high proportion of general agricultural subsidies go to very large producers. The distribution of commodity subsidies might be so skewed as to argue in favor of limitations on payments to wealthy farmers and greater support for the more widely distributed conservation funds.

Since 2001, the Environmental Working Group (EWG) has posted on the internet an updated database of USDA agricultural payments, searchable at the producer or farm level. Payments can be aggregated for counties, congressional districts, or states. It shows how high and how concentrated payments are in commodity subsidy programs, such as cotton or rice, and how widely distributed, in contrast, the rental contract payments are in the CRP. In fact, in its first year of existence, EWG's farm subsidy database was searched more than 60 million times, as congressional staffers, interest groups, and journalists tried to analyze farm policies and farmers tried to see how they and others were faring under the subsidy programs.[3]

Theories about congressional delegation of discretion to federal agencies often point out that third parties, such as interest groups, can end up monitoring agency decisions. These interest groups take the data generated by agency decisions and condense it in studies or make it available in raw form to others. EWG is the archetype of information-based interest groups. Founded in 1993 by Ken Cook and Richard Wiles, EWG has steadily influenced the debate about

U.S. farm policy by liberating data from USDA and bringing it into congressional deliberation. In the mid-1990s, this involved obtaining data stored on computer tapes from USDA and developing reports to show Congress certain patterns that it did not highlight in its reports about the CRP. Once the internet became a lobbying tool, the EWG used the Freedom of Information Act to get data on individual payments under farm programs and post this information on its website.

The information provided by EWG has changed the terms of many debates about farm policy. Early on, EWG analyses provided the basis for discussion in congressional hearings. The expansion of the data into a searchable database on the internet identified the EWG website to staff, interest groups, and constituents as a place where they could find out answers to one of the basic questions in politics: who is getting what from whom toward what end? As information technology changed, so did the approach of EWG. Originally, it provided analyses through hard-copy studies and press releases circulated at key points in farm bill debates. It evolved to providing a searchable internet database on farm payments, discussing policy developments on a farm policy blog (appropriately entitled *Mulch*), and even posting videos of its press conferences on *YouTube.com*. The local nature of the EWG data has attracted the attention of newspaper reporters writing local stories about the CRP and the notice of members of Congress interested in who is benefiting from agricultural payments in their districts.

EWG, a 501(c)(3) nonprofit organization, explicitly states that "the mission of the Environmental Working Group is to use the power of public information to protect public health and the environment." For more direct political work, the group created the EWG Action Fund to advocate for "health protective and subsidy-shifting policies." The group works on a number of topics, including providing safe drinking water, reducing toxins in consumer products, and shifting agricultural subsidies away from commodity programs and toward conservation expenditures. In each of these areas, however, the way that EWG seeks to bring about change is through information provision. The language of its website shows how it is able to capture and redirect attention in political battles: "Our research brings to light unsettling facts that you have a right to know. It shames and shakes up polluters and their lobbyists. It rattles politicians and shapes policy. It persuades bureaucracies to rethink science and strengthen regulation. It provides practical information you can use to protect your family and community. And because our investigations and interactive websites tend to make news, you've probably heard about them. Even if you've never heard of us. Which is fine. We'd rather you remember our work than our name."[4]

This chapter examines one part of EWG's policy efforts—the use of information provision to change the debate about the operation of and funding for the CRP.

INFORMATION 1.0

In February 1994, EWG released a 38-page analysis of the CRP entitled, "So Long, CRP." The report, based on computer analysis of 375,000 contract records that EWG received from USDA, argued that "conservationists need to 'protect their base' in the 1995 Farm Bill" by devoting attention to the fate of acres set to expire from CRP protection.[5] It also noted that the CRP had grown to more than 20% of agricultural assistance since 1989, with an overall payment of almost $20 billion generated by the 36 million acres under contract. The EWG study predicted that "farm programs that offer both economic assistance to farmers *and* significant environmental benefits and incentives will attract a wide and solid base of political support."[6] The report also recommended that USDA conduct a broad analysis of environmental risks in agriculture. Cook, the president of EWG, indicated that if USDA distributed funds based on comparative risk assessments, then "conservation funds would shift *from* areas where wind erosion is the primary problem *to* areas where water quality, wetlands protection, and worker health risks are the primary problems.[7]

In August 1994, Cook testified before the House Agriculture Subcommittee on Environment, Credit, and Rural Development in a hearing entitled, "Review of the Budget and Policy Consequences of Extending the Conservation Reserve Program." To bring home the message of the report, Cook provided the committee members with an appendix to his testimony that gave each subcommittee member a one-page, detailed description of the CRP contracts in each member's district (e.g., the total number of contracts, acres enrolled, CRP payments, and tons of soil saved annually by the CRP). Cook noted at the start of his appearance that this appendix was not a random sample of congressional districts. He estimated that nearly half of all CRP funds went to contracts in the districts of the subcommittee members and that seven of the top 10 districts receiving CRP funds were represented by members of the subcommittee.

The detailed data were a hit with the committee members. As Representative Combest (Texas) put it, "Mr. Cook, this information, this is some really interesting stuff you put together here...This shows us some things here that we didn't know before. I think we ought to start figuring the payments based upon tons of soil saved. We are going to come out like a bandit if we do that in west Texas."[8]

Representative Sam Farr (California) was similarly appreciative of EWG's analysis of CRP data for his district, noting,

> I want to thank Mr. Cook for his—when I came in here, I didn't know how much land of my district was in the program, and now I have it down to the minutia, and I really appreciate it. In fact, what I am going to do is go back and apply what I have been saying here, what I have heard today, to these particular pieces of land and see if I can come back to this committee with some constructive recommendations for how to improve the program as we go into the farm bill next year.[9]

During questioning, Cook explained that EWG had conducted the "So Long, CRP" study, which analyzed program information on computer tapes received from USDA (from one of its mainframe computers), in part because various constraints kept USDA from doing this type of analysis. Representatives on the subcommittee were surprised that an interest group, rather than a department or agency, had to perform this type of program analysis. The exchange between Representative Peterson and Cook captured a sense of exasperation.

> *Peterson*: Mr. Cook, I want to commend you for doing this analysis...I would somewhat chastise the Department [of agriculture]. It seems kind of ludicrous to me that somebody such as yourself and your organization would have to do this, that the Department would not be able to do this...How did you get this information from the system? Did they give it to you on computer tape?
>
> *Cook*: On main-frame computer tape. We have similar records, actually for all ASCS programs [e.g., farm programs] now—in more detail.
>
> *Peterson*: You feed it into your PCs and analyze it. But they [USDA] don't do this, do they?

Cook: I think it varies from place to place within the Department. Some seem to be a little further along than the others in making information readily available outside the experts who run the main frame computers.

Peterson: But the people in the Department don't have that information in the form you have it because they have not done the work to get it in that shape, right? If they did have it you would not have had to go through all this work, right?

Cook: I guess that is right...we have made the information available without charge to the Department.[10]

The interest of the members of Congress at the oversight hearing in EWG's analysis of the CRP's impact in their districts shows two of the advantages the interest group used in its work. EWG became famous for liberating data from USDA and putting together analyses the agency could perform, but chose not to do. The second signature of EWG was to make the analysis local, which attracted the attention of local media. In describing how it conveyed the results of its 1994 analysis of the CRP, Cook said that "the Environmental Working Group has disseminated literally thousands of analyses of CRP contracts from our database, providing data at the county, congressional district, state, and national levels."[11]

The local angle generated media coverage in local newspapers. In an April 1994 *St Louis Post-Dispatch* article, a Washington, DC-based reporter used the EWG information to show the impact of the program in Missouri. The article, "Farmers Face Loss of Green Payments," opened with these findings: "Missouri farmers stand to lose $1.16 billion in conservation grants—among the highest in the country—unless the federal government devises a system of 'green payments' to ease budget cuts, an environmental group has concluded. Illinois farmers could lose $668 million. In a computer study, the Environmental Working Group of Washington has reported that the government's plan to phase out the Conservation Reserve Program beginning next year could take a heavy toll on farmers and on millions of acres of protected land."[12]

When EWG produced its 1994 report—"Blowin' in the Wind: How Much Soil did the CRP Save?"—the local figures generated by the group again attracted attention from local media. An *Omaha World Herald* article on the study indicated that "in Nebraska...the federal program to halt erosion on 1.4 million acres of fragile cropland will save less than half the amount of soil that the U.S. Department of Agriculture had said it would, the Environmental Working Group said."[13] The report emphasized a common theme in the 1994

EWG statements and analyses, that the CRP could be better targeted to focus on "water quality, wildlife, endangered species and wetlands."[14]

Eventually, USDA adopted a transparently defined Environmental Benefits Index that did score lands in the CRP bidding process on dimensions that groups, such as EWG, had recommended. When the secretary of agriculture announced in May 1997 the signup of a new set of lands in the CRP based on these broader environmental benefits, Cook used the 1990s version of support—a written statement—to provide praise for this action. Cook declared, "The 16.1 million-acre enrollment in the Conservation Reserve Program (CRP) announced by Secretary Glickman today is the most important single action the federal government has ever taken to help farmers and ranchers conserve resources and protect the environment. This is a milestone in conservation policy. With a stroke of his pen, Secretary Glickman has done more to conserve wildlife on private lands than any government official before him."[15]

INFORMATION 2.0

By the time the 2002 Farm Bill was debated, main frame computers had given way to desk top PCs and the wide availability of information on the internet. EWG changed its information provision tactics to keep pace with the evolution of information processing. As Congress debated changes in agricultural policy, EWG posted a searchable database on its website, the results of years of data requests under the Freedom of Information Act and intensive website development.

Cook credited the database with changing the information base that people had during farm bill debates and changing the rhetoric that legislators and lobbyists used in discussing farm policy. He noted that there were thousands of articles that mentioned the EWG database in coverage of farm policy. Breaking down why he believed the database had such an impact, he noted three effects. First, it showed how much of the subsidies flowed to large producers: "It demolished the symbolism of farm subsidies as the salvation of family farms. It repainted Grant Woods' 'American Gothic' as a political cartoon. You just can't hand a pitchfork to a collection of millionaires and Fortune 500 companies and state prisons and universities and other unlikely recipients, or drape them in a dress or overalls, and pass them off as family farmers."[16]

Second, it highlighted how many farmers received little or no help from government payments: "You can't say farm subsidies are saving the family farm when most farmers get no subsidies. They grow the wrong crops or raise the wrong livestock. And those who get nothing really noticed that, and what their neighbors got, too."[17] Third, people saw how much of the farm payments flowed to commodity subsidies rather than conservation: "And all of the farmers who are turned away from conservation programs year after year for lack of funds noticed how subsidies didn't work for them."[18]

EWG's detailed data on the large subsidy payments flowing to large agricultural producers helped spur debate in 2002 about placing more stringent limits on the amount of agricultural payments a person might receive. The public focus on the flow to large producers created lobbying problems for some interest groups whose members received relatively large subsidy payments. As Kenneth Hood, the chair of the National Cotton Council, put it, "At the core of our dilemma is finding a way to make the public feel good about six- and seven-figure payments to farmers."[19] Cook pointed out that the attention to payment levels changed the public debate about the farm bills, although the private resolution of the legislation's terms in conference committee was ultimately less targeted toward reform. Describing the resolution of the 2002 Farm Bill to a group of agricultural bankers in 2004, Cook recalled,

> The problem as Mr. Hood saw it, was that for the first time, the public knew about those payments in embarrassing detail. Worse yet, from the standpoint of the Cotton Council, the Rice Growers and the Farm Bureau, political leaders in both parties, including many senior farm policy makers, were moved to act by this very same information. It was only behind the closed doors of the 2002 Farm Subsidy Bill conference committee that the big growers' problems disappear. A working majority of Democrats and Republicans agreed, dishonorably in my view, to go against the positions of the House and Senate. They removed the payment limits and reduced conservation spending below a fair compromise between the two positions...It continued the flow of taxpayers' money, unimpeded, to the nation's largest producers—10% of whom collect over 70% of the subsidies. Mr. Hood, like many in agriculture, blamed the close call on the payment limits, the intense debate over farm subsidies in general, on me, and my colleagues.[20]

As EWG updated its farm subsidy database, it would issue press releases noting the new figures available on the website. A press release in November 2005 noted that over the last decade 22 congressional districts collected more than half of all agricultural subsidies.[21] The top 10% of recipients received more than 70% of the farm subsidies (i.e., $104 billion out of $143 billion in total subsidies). EWG's strategy focused on the disparities in commodity payment programs partly in an attempt to shift more funds and attention to conservation programs in agriculture, such as the CRP. Cook explained to *Audubon Magazine* in 2005, "There's no question spending more money on conservation is a better deal for taxpayers than price supports for things we already have too much of. We would much rather see taxpayers' money going into helping farmers with how they farm as opposed to paying them to grow more of what we don't need."[22]

When the Farm Bill came up for debate in 2007, EWG was ready with another innovation in its subsidy database. With new information from USDA, EWG was able to assign subsidy payments more accurately to individuals. Previous versions sometimes listed large farm operations as the recipient of subsidies, although the ownership might have been spread across many individuals. This corporate structure was by design. As EWG put it, "these very large operations included multiple owners who divided farms into multiple businesses in order to have enough individuals to maximize per-person federal farm subsidy payments, which are much higher per acre for cotton and rice, while complying with various per person payment limitations."[23]

As the debate over limitations in payments heated up, some participants proposed means testing for commodity payments, which would stop subsidies flowing to those with adjusted gross incomes greater than $200,000. The evolution of Web 2.0 tools meant that Cook could provide real-time comments on the debate through his postings on EWG's agricultural blog *Mulch*. In a May 2008 posting, Cook pointed out that subsidy lobbyists were trying to scare conservation groups away from reform proposals dealing with payment limits by suggesting the limits should also be applied to conservation payments. Cook noted that EWG took a consistent position on this: "means testing is clearly appropriate for commodity subsidies. Taxpayers should not be accelerating farm consolidation by bankrolling big, wealthy farm operations, which is what commodity subsidies do. But with tight funding and such a large, unmet need for conservation programs, we should also take a hard look at means testing for all farm bill programs, even conservation."[24]

The informal nature of blogging meant that Cook could be much more explicit about his take on political battles than in the earlier era of statements delivered via press releases. In a *Mulch* posting, he said that subsidy lobbyists had an unsubtle message for conservation groups.

> The message is always the same: play nice, keep your great big save-the-world pie hole shut about Title 1 [commodity subsidy] reform, and maybe you'll get some money. A lot less money than is needed, a lot less money than you want, a lot less money than you've told your supporters you'll be fighting for in the Farm Bill, of course. But if you play along, the commodity boys *will* throw you a few bucks that are mandatory (and many more bucks that are merely authorized) for programs you consider central to your mission, whether it is holding the nation's migratory bird flyways together, protecting farmland from sprawl, preserving ecological gems or helping family farms protect their land and everyone's water, air, and wildlife. Big Ag will even consider "protecting conservation through conference" if you bend sufficiently to their wishes, though of course you never know what the House (or Senate) might do… some things are beyond their control…Blah, blah, blah.[25]

The Environmental Working Group allows public comments on its blog, which means people can post their reactions to the data the group have extracted from USDA. Most of the comments about the subsidy database are positive and point out the benefits of knowing the details of these payments. A farmer listed in the database, however, offered a different perspective. He was particularly upset that EWG linked the issues of poverty and farm subsidies by placing information on local income distribution next to information about agricultural payments.

> You forgot to state that also in this new database is the average income in the area and how many children are in poverty. You have it on my database page. I am a 25-year-old farmer that started farming corn and soybeans at age 19. Not with Dad or family or anyone. I have an established business now, but I own no real estate. Equipment payments consume most of my income. My equipment loan would be the same as a new business start-up loan to all you non-farm people. I do not live a life of luxury. I work hard to keep the lifestyle I have. Can you please put on my database page that I employ the father of two poverty-stricken kids? I give him a paycheck, food, money for the kids' Christmas. I also can't control where that paycheck gets spent. I also utilize no-till practices and grow food-grade white corn. Yes, remember farmers grow food, not grocery stores. So, if you think I make $50K a year and ignore the 5,000 kids in poverty in my area, you are wrong. So, could you please post the

information I gave you next to my name in your database? I don't care if the world knows how much subsidy payments I have received. I just don't like the personal attack on every subsidy recipient you have done.[26]

EWG got its message of reducing subsidies and increasing conversation expenditures out in numerous ways in 2007. When Cook addressed the National Press Club in Washington, DC, in June 2007 to mark the release of a new version of the subsidy database, a (lightly viewed) video excerpt ended up posted on *YouTube.com*. When the *New York Times* wrote in February 2007 about the Bush administration's proposal to reform farm policy by reducing payments to farmers by $10 billion and prohibiting payments to producers earning more than $200,000 in adjusted gross income per year, Cook was quoted as saying the secretary of agriculture had "really planted a flag for reform" and called it "the most detailed proposal I have seen in the 30 years I have been working on agricultural issues."[27] Articles in local newspapers in 2007 used the EWG database to provide a local angle on agricultural payments. The *San Francisco Chronicle*, for example, found through the EWG data that a wealthy brother and sister in San Francisco received $2.4 million from cotton subsidies in 2003–2005.[28] The article even had a human interest angle, quoting one apparently wealthy farm operator who was not wildly supportive of the program as saying, "you work hard for your paycheck every week. Why the hell should you give part of it to me?"[29]

After the Farm Bill passed in 2008, EWG continued to use the internet to post follow-ups on how conservation measures fared in the appropriations process. The EWG website provided detailed information for environmentalists and others to track how legislators lived up to promises made in the bill. EWG's language was again direct and colorful, with a September 2008 update noting,

> behind the thin green gloss congressional leaders spread across the subsidy-laden 2008 Farm Bill, key Democratic lawmakers are hacking away at promises to expand conservation and other environmental programs...[W]ithin weeks of the Farm Bill's passage, the Senate Appropriations Committee sent to the Senate floor a spending bill (S.3289) that would slash conservation measures by $331 million in fiscal year 2009...[T]his farm bill bait-and-switch routine by the Democratic Congress mirrors a longstanding Republican tradition of broken promises where pledges to increase money for environmental programs are followed by systematic and dramatic cuts that have left conservation programs billions short over the past decade.[30]

This legislative update noted that the recommended 2009 appropriation of almost $1.9 billion for the CRP was the same as the level pledged in the farm bill. But, the Appropriations Committee recommended levels of funding lower than those promised in the farm bill for such conservation programs as the Environmental Quality Incentive Program, the Wildlife Habitat Incentive Program, the Grazing Land Reserve Program, and the Farmland and Ranchland Protection Program.

Overall, EWG has pursued a consistent and successful strategy of taking information collected by USDA, organizing and analyzing it to provide legislators, journalists, and the public with a detailed understanding of programs like the CRP. In the mid-1990s, this meant running computer analyses on mainframes to produce reports that would be distributed to congressional staffers and reporters. Today it means a widely searched and accessible database on the internet, supplemented by blog postings and *YouTube* videos. In adding up its successes in farm policy debates, the EWG website listed these achievements:

> Congress voted to increase conservation payments to farmers by 80%, providing $13 billion over six years for programs to reduce water contamination and soil erosion, as well as to protect wildlife. EWG's searchable Farm Subsidy Database, which tracks recipients of 90 million government agriculture-program checks, has shaken up the debate over continuing wasteful subsidies to big agribusiness: "...President Bush proposed drastic cuts to wasteful farm aid in his 2006 budget. Data compiled online since 2001 by EWG—and the media attention generated through our advocacy—put public pressure on the [Bush] administration, resulting in a proposal to cut farm subsidy payments by 5% and close loopholes that allow farmers to pocket $1 million or more each year."[31]

Another set of observers shared EWG's belief in the impact of its strategy to provide data—the journalists who drew upon the EWG data and analyses. The accolades for EWG are widespread. Reporters have described EWG as "an environmental group with clout" (*USA Today*), "a green dream team of computer programmers, policy experts, and engineers" (*Associated Press*), and "an environmental advocacy organization that has a knack for shaping government data into punchy calculations" (*Atlanta Journal-Constitution*). News reports during farm bill debates regularly credit EWG. A 2007 *New York Times* piece noted, "Thanks to the Environmental Working Group, we know exactly how much money every subsidized farmer is getting in every county.

The group's database shows that just 1 percent of all farmers receive about 17 percent of the payments—averaging $377,484 per person, over three years." In 2005 the *Des Moines Register* concluded that the "Environmental Working Group's internet database was influential in the debate over the 2002 Farm Bill and almost certainly has increased support in Congress for tightening limits on subsidies to large farms." The *New York Times* in February 2002 described the course of the farm bill debate: "throughout the angry Senate debate about whether to limit subsidies to wealthy farmers, lawmakers kept referring to 'the web site.' No one had to ask, 'what web site?'—it was www.ewg.org, operated by the Environmental Working Group, a small nonprofit organization with the simple idea that taxpayers who underwrite $20 billion in farm subsidies have the right to know who gets the money."[32]

CHAPTER 6

MEDIA COVERAGE AND ACADEMIC ANALYSES: CYCLES OF PRAISE AND CRITICISM

Journalists and researchers operate in very different information markets, which partly explains the different perspectives that these observers offer on the operation of the Conservation Reserve Program. The media stories that emerge about the CRP are determined strongly by demands from particular audiences. Newspapers satisfy at least four different types of information demand: producers/workers, consumers, entertainment seekers, and voters. Looking at newspaper coverage of the CRP, one can see articles that fit all four categories.

In farm states, stories about the operation of the CRP are local business stories. Articles may mention how to sign up with the Farm Service Agency and when enrollment bid deadlines are approaching. Consumer stories about the CRP relate primarily to hunting, with articles describing how the CRP impacts local hunting and indicating where and when hunting on CRP lands begins each season. Sometimes a human interest angle will result in a CRP story aimed at the entertainment demand. The need to be diverted and informed, however, is more influential in the presentation of political stories about the CRP. The majority of newspaper articles about the CRP deal with the political battles around the renewal of a given farm bill. Local or regional newspapers rarely write about the details of how the CRP actually is implemented. It is extremely rare to even see the words "Environmental Benefits Index" in print. Voters' "rational ignorance" about the details of policy means they do not demand stories about the precise workings of the CRP. Yet, the often-dramatic legislative battles do generate coverage.

The many articles about the debate in Congress often have a horse-race flavor, focusing on who is ahead and who is behind in reaching their legislative goals. The CRP is discussed as a contest between the president and Congress, the House and the Senate, farm states and urban areas, and among interest groups, such as farmers, hunters, environmentalists, and agribusiness companies. In Washington, DC, people have a producer demand for political information because politics (in the form of government employment or policy advocacy) can be a person's day job. This means that the *Washington Post* is more likely to cover the progress of a farm bill through stages of committee hearings, floor debates, conference reports, and presidential signature. The *New York Times* also provides, for its targeted national audience of college-educated readers, coverage of the debates over the billions of dollars involved in farm policy.

The focus on the CRP during the renewal of farm legislation means that stories will carry both criticism and praise, highlighting quotations from the multiple parties involved and statements by CRP proponents and opponents. Supporters of the CRP likely to be quoted include representatives of farmers, environmentalists, and sports groups (e.g., hunters). Opponents include agribusiness processors who pay higher commodity prices because CRP restricts cropland production, farming input suppliers who lose business when croplands go into conservation, and legislators or administration officials focused on the impact of expenditures on taxpayers and the federal deficit.

The origin of news stories sometimes comes from the supply side. During periods of legislative debate, interest groups anticipate that reporters in farm states and in national papers may want to write about the farming debate. These groups often produce reports and hold press briefings to lower the cost of congressional staffers and journalists learning about their particular perspectives on the CRP. Legislators, often depicted in political economy models as credit claimers and blame shifters, may write to their local newspapers during and after passage of the bill to take credit for provisions that benefit their constituents. The secretary of agriculture also figures in these stories, at times by traveling to a farm state to hold an event that symbolizes progress in the evolution of the CRP.

Media stories about the CRP center primarily on political economy, who gets what from the program, and why a particular group's lobbying and advocacy is successful. The focus in the academic literature about the program is much different. Most of the articles are published in natural science journals and discuss soil, agricultural production, or wildlife. Social science articles look at particular questions of agricultural economics or the economics of environmental protection and preservation. Although most of the authors come from the nonprofit sectors of academia or government, the types of questions pursued reflect the demands of the policy analysis market and relatively narrow questions that involve precisely stated hypotheses that are then tested through empirical work. What is measured and tracked often determines what can be asked and investigated. Trends in academic research topics, such as the degree of rationality of producer decisions, the nature of nonmarket valuation, and the distribution of benefits by race and income, also show up in articles about the CRP. The preference for data and statistical analysis of past decisions means that most of the academic work on the CRP looks backward at previous implementation and is less likely to offer predictions about the likelihood or desirability of future policy scenarios.

MEDIA COVERAGE OF THE CONSERVATION RESERVE PROGRAM

In the initial years of the CRP, newspapers with brand reputations for strong, hard news coverage and business outlets covered changes in the implementation of the program. In the *Journal of Commerce*, a publication aimed at business readers, an article in 1987 reported how farmers were rushing to sign up in the CRP due to a one-time USDA bonus for producers willing to sign up corn cropland (a boost to Midwestern farms). Deputy Secretary of

Agriculture Peter Meyers indicated that, since the bonus for corn cropland might not be repeated, "my advice to the farmers would be to run, not walk, to your nearest ASCS office and sign up this time, especially if you're going to put corn base [acreage] in."[1]

The announcement by Agriculture Secretary Richard Lyng that 19.5 million acres were enrolled in the CRP as of March 1987 prompted articles stressing that the program had enrolled acres more quickly than anticipated (to meet a goal of 40 million acres by 1990). The subheadings of the *New York Times* article "Anti-erosion Program for Farms Accelerates" show the balance of praise and blame often featured in mainstream newspapers: "Nearly Halfway to Goal," "Well Ahead of Schedule," "Fears of Negative Effects," and "Concern about Outsiders." This article emphasized that taking cropland out of production would reduce erosion and pesticide runoff into streams. The balancing quote of criticism came from Phillip Hoffman—a farm supplier and grain merchant who worried that the program took money out of circulation in local farm economies because fewer inputs were bought and the CRP checks went to producers who did not live locally—who noted, "It's a positive force for erosion control. But it has got out of hand. Too many acres are being signed up here in the Green Hills, and too much of the money is going to absentee landlords. This area will never see a lot of that money. It's going to be very detrimental."[2]

The *Washington Post* also charted the changes in the CRP's operation early on. In June 1988, for example, a *Post* headline announced, "White House Forms Interagency Panel to Devise Plan to Cope with Drought."[3] The piece noted that Agriculture Secretary Lyng had authorized producers to allow hay harvesting (but not livestock grazing) in counties that were severely affected by the drought.

As more farmers signed up for the CRP, more regional and local newspapers began to cover the program, often with a local angle. The *St. Louis Post-Dispatch* ran a five-part series on the CRP in May 1989, concentrating on Missouri farmers. The leads chosen for these articles show how the papers had to make these stories entertaining for local readers who might not naturally seek out policy-related stories. The series opened with these lines: "Missouri farmers not only are having dirty thoughts, they're getting paid for them. They're thinking about dirt and how to keep it where it belongs. The payment is from the Conservation Reserve Program, an ambitious federal soil conservation program started in 1986, after passage of the 1985 Farm Bill."[4]

The fifth article in the series had a similarly entertaining hook: "What's the Incentive? Actors call it 'motivation.' 'What's my motivation for doing this?' they ask. Kids call it 'Why me?' especially if it is something like mowing the lawn. Landowners call it 'incentive.' Why should a landowner put his acreage in the Conservation Reserve Program? For the farmer who will lose his farm if he doesn't, the answer is simple. Live or die. But for others, it is more complex."[5]

The five-part series gave readers a great deal of local information about the CRP, noting that the average bid was $65 per acre in northern Missouri, that the state conservation department was studying the impact of the CRP on wildlife habitats, and that farmers could receive additional money from the state to implement their conservation plans if the cover planted benefited wildlife.

As the debate over the 1995 Farm Bill began, reporters in farm states who covered hunting began to focus on the benefits of the CRP. In a March 1994 article in the Minneapolis, MN, *Star Tribune*, "CRP Isn't Just for Farmers; Renewal of program to benefit wildlife, environment," staff writer Dennis Anderson argued in the sports section that wildlife benefits had been significant from this program: "Minnesotans also should remember that it makes little sense to fund ambitious environmental programs, such as the cleanup of the Minnesota River, while not making a strong case in Washington for full renewal of CRP. Without CRP, the Minnesota [River], polluted as it is, will suffer further. If you want government to spend your tax dollars wisely, contact your congressional representatives. Tell them you're concerned about CRP, and you want it renewed."[6]

Article titles in the *Star Tribune* sports/outdoor section later that year made a similar point, such as the headline and lead reading, "Thanks to CRP the Good Old Days Are Back; 1994 Pheasant Opener: A wildlife blessing, the Conservation Reserve Program is at a crossroads. In the balance hang wildlife, tons of topsoil and clean water."[7] Sometimes the CRP simply emerged in references in hunting articles, as in the 1995 article in the *Omaha World Herald* that began: "Two pickup trucks slowly cruised past a Conservation Reserve Program field that was planted with tall grasses. 'I'll bet there are 100 birds in that field,' said Dan Snider. 'But we're not even going to hunt it. You close the door on the truck and they'll be gone. They're really spooky in that tall grass, especially at this time of the year.'"[8]

Regional papers in farm states often covered the political debates in Washington when there was a local angle. The Minneapolis *Star Tribune* ran an article in 1994, entitled "Congress Is Urged to Spare the CRP; Plan cited as benefiting wildlife, improving soil, lifting grain prices," when Minnesota and North Dakota conservationists went to Washington to brief 20 members of the Congressional Sportsmen's Caucus and their staff about the environmental and economic benefits of the policy.[9] The *Omaha World Herald* covered the congressional testimony of an Omaha agribusiness executive when he went to Washington to argue that programs, such as the CRP, made U.S. farmers "less productive, less efficient, and less competitive."[10] The article title "Ag Processor Argues against Set-Asides" neatly summarized the position of those whose businesses suffered when croplands were idled, commodity prices increased, and purchases of farm inputs, such as fertilizer, dropped. The *Omaha World Herald* also made sure it covered changes in policy through the eyes of their local representatives. When the Clinton administration announced support for extension of the CRP, the paper reported these reactions: "'I am pleased the uncertainty that has surrounded this program has been removed as we get ready to write the 1995 farm bill,' said Sen. Bob Kerrey, D-Neb., a member of the Senate Agriculture Committee. 'I am very pleased that the U.S. Department of Agriculture has basically adopted my proposal,' said Rep. Doug Bereuter, R-Neb."[11]

Papers provided local links and statistics as they wrote about the political battle that ensued over the 1995 Farm Bill. In the article "Cropland Conservation Program Left in Limbo; Farm-bill negotiations could set its course," the Minneapolis *Star Tribune* began an update on DC policymaking with this lead: "Few are watching the nation's Capitol with keener interest these days than the thousands of Minnesotans who agreed to idle their farmland for 10 years in return for $100 million in annual 'rent' checks from Uncle Sam. The House and Senate, in passing markedly different farm bills, have left in limbo the future of these long-term deals, arranged under the Agriculture Department's Conservation Reserve Program."[12]

The media coverage that greeted President Clinton's signing of the Farm Bill in 1996 provided readers with political summaries that credited both environmentalists and hunters for the passage of the CRP. As an April 1996 article, "Outdoors Folks Deserve Credit for Farm Bill," put it,

> "The effort put forth by individual sportsmen, contacting their congressmen, really made a difference," said Lloyd Jones, field director for the Delta Waterfowl Foundation, who was involved in the CRP battle. "There's no question that the whole information and education effort was the key factor on how Congress was able to continue the programs...Part of that effort was the understanding that this was a program that provides a lot of benefits to a lot of people, as opposed to the traditional commodity price supports, which basically benefit only the farmer or landowner involved."[13]

A consortium of conservation groups joined the effort. Sports-based organizations, such as the Wildlife Management Institute, the International Association of Fish and Wildlife Agencies, the National Wildlife Federation, Delta Waterfowl, and Ducks Unlimited, led the way. They were joined by mainstream environmental groups, such as Sierra Club, the Audubon Society, and the Natural Resources Defense Fund, among others.[14]

After the Farm Bill passed and additional signups for the CRP began in March 1997, coverage in local newspapers with farm readership gave details on how to enroll. In a March 1997 article, "Deadline Nears for Farmers to Reap Benefits of Program," the *Atlanta Journal-Constitution* provided this advice: "In order to be approved for the program, applicants must meet several requirements, including ownership or operation of the farmland for at least one year. And the land must have been used for crops for two years between 1992 and 1996. Southside landowners interested in participating in the program can call or drop by the Farm Service Agency for their area."[15]

The Minneapolis *Star Tribune* article, "Farmers Flock to CRP, but Not All Will Qualify," noted that USDA would use a scoring system (the Environmental Benefits Index) to encourage greater environmental benefits from the enrolled lands. The article quoted Kevin Lines, the farmland program leader for Minnesota's Department of Natural Resource: "Lines said that landowners wanting to renew their CRP contracts likely will have to be willing to reseed their lands to enhance wildlife habitat in order to score enough points to qualify for the new program. Lands nationally will be graded by an Environmental Benefits Index, and without enhancement, many of Minnesota's existing CRP acres might not qualify, Lines said [*sic*]. 'The status quo isn't going to be sufficient,' Lines said."[16]

The coverage by the *Omaha World Herald* of the 1997 CRP signup shows that, in states where the CRP was important to incomes, the stories were highly detailed. In an article title unlikely to pique the interest of suburban readers, the paper ran an assessment headlined "Environmental Benefits Index Now Rates CRP Land." The paper summarized the likely impacts of the new scoring system: "Farmers who offered to place their land in CRP during the new sign-up heard a new phrase—environmental benefits index. Most of the emphasis during previous sign-up periods was placed on the prevention of soil erosion. But the benefit of CRP to wildlife now was evident. Organizations such as Pheasants Forever and Ducks Unlimited sold CRP to Congress under that concept. A scoring procedure, based on the environmental benefits index, was developed, and land that offered little or no benefit to wildlife was rejected."[17]

With the CRP heavily focused on environmental benefits, political stories about the policy emphasized this aspect of the program. In the third presidential debate in 2000, Vice President Al Gore probably caused the most people at one time to notice (briefly) the policy when he said, "I think that we ought to have an expanded Conservation Reserve Program. And I think that the environmental benefits that come from sound management of the land ought to represent a new way for farmers to get some income that will enable them—enable you—to make sensible choices in crop rotation and when you leave the land fallow and the rest."[18]

When President Bush signed the 2002 Farm Bill, media coverage noted how the measure expanded the CRP ceiling from 36.4 million acres to 39.2 million acres. The first line of the *Rocky Mountain News* article about the bill's passage stressed that "the 2002 Farm Bill increases spending for wildlife conservation by 80 percent from the 1996 version and raises the ceiling of land that can be set aside under the Conservation Reserve Program by nearly 3 million acres."[19] President Bush highlighted his support for the CRP during the 2004 presidential campaign, in part by visiting the Katzenmeyer Farm in Le Sueur, MN, on the same day in August 2004 that he announced three new efforts designed to strengthen the CRP, including a proposal to allow early re-enrollments and extensions of CRP contracts.[20]

Stories about the CRP also continued to show up in sports pages. In 2003, Nebraska Governor Mike Johanns praised the efforts of the Nebraska Game and Parks Commission to join with the group Pheasants Forever to create a plan—"Focus on Pheasants"—to improve habitat on lands owned by the state and lands enrolled in the CRP. As Johanns put it, "State identity, in my opinion, is football and pheasants. Pheasants are truly a part of who we are."[21] Howard Vincent, chief executive officer of Pheasants Forever, stressed that his group's lobbying for the CRP was essential to its mission. In an interview with the Minneapolis *Star Tribune* in 2004, Vincent noted that the group had bought 19,442 acres in Minnesota and turned the land over to the state for hunting. He warned that incremental additions might appear small, "Yes, adding a 400-acre parcel is miniscule. But go out to one of our chapters and ask them if that 100-acre parcel that they bought down the road is meaningful. They can see pheasants on it."[22] Vincent also noted that the group's lobbying efforts were very effective in having broader impacts on the environment and remarked, "With the stroke of a pen, we can get more done than we could with 50 years of banquet dollars. That's 36 million acres [of CRP nationally]."[23]

The debate over the 2008 Farm Bill was especially contentious. In an era of relatively high commodity prices, businesses (such as food processors) supported reducing conservation programs in the hope that increased croplands would lead to decreased commodity prices. The political battles over conservation attracted heavy attention in local and national coverage. April 2008 headlines in California included "Farm Bill Fight Intensifies: Conservation Funds Now Focus of Finalizing Deal" in the *Stockton Record* and "[U.S. Rep.] Herger Thinks Cutting Conservation Funds in New Farm Bill Is All Right."[24] The *New York Times* noted the attempts to change the system in its article "A Bid to Overhaul a Farm Bill Yields Subtle Changes," which quoted Senate Agriculture Committee Chairman Tom Harkin (Iowa) as saying, "We have to consider new ideas. We should not cling to a system that channels ever larger commodity payments to a relatively few, with two-thirds of American farmers getting none at all."[25] In an opinion column published in the *Bismarck Tribune* in January 2008, Senator Kent Conrad (North Dakota) made media coverage part of the political debate: "We're also fighting against an East Coast media that simply doesn't understand farming and is encouraging opposition to the farm bill. They overlook the fact that this is more than a farm bill. It is a 'food bill.' Sixty-six percent of the money in this bill is for nutrition programs to help feed those less fortunate, as well as our nation's children."[26]

President Bush vetoed the Food, Conservation, and Energy Act of 2008, in part arguing that the bill did not do enough to limit federal payments to wealthy farmers. Congress overrode the veto, which meant new provisions governing the operation of the CRP were enacted. Conservation spending accounted for about 9% of the bill's total spending, with an authorization for the government to spend $7.9 billion more than current conservation spending levels. Overall, the maximum number of acres enrolled in the CRP was reduced to 32 million acres. Farmers with adjusted gross incomes of more than $1 million would, under some circumstances, not be eligible to get farmland conservation payments.[27]

What was new in the 2008 Farm Bill debate was the increasing ability of information about farm policy made possible by blogs and interest group websites. This led to more immediate circulation of information and the existence, for those inclined to seek it, of more personal and diverse opinions expressed about USDA's conservation programs. Consider these three different perspectives on the passage of the 2008 act. At the Birding Business News website in June 2008, the article "Farm Bill News from the Bird Community E-Bulletin" said:

> The mainstream media watched the House and Senate pass the bill in early May, only to have it vetoed by President Bush, and then overridden by Congress. Most of the media's focus was on the level of subsidies to large farmers, the perception (and reality) of "pork," a new "permanent disaster" program, and nutrition elements. Conservation elements within the Farm Bill were given little serious attention. That was unfortunate, since the status of conservation features of the Farm Bill is particularly important for grassland and wetland birds and other wildlife. At the end of this process, the conservation elements for birds were mixed.[28]

Jesse's Hunting and Outdoors website updated sport enthusiasts about the progress of the bill, noting that 20 of the 26 Farm Bill Conference Committee legislators belonged to the Congressional Sportsmen's Caucus. The web posting "Congressional Sportsmen's Caucus Preserves Funding" reported that "thanks to sportsmen-legislators, hunters in rural America can look forward to local farmers continuing to set aside acres of habitat for wildlife."[29] At *Redstaterebels.org,* ecologist George Wuerthner criticized the CRP as a "Boondoggle in the Fields." He acknowledged that the CRP did provide some environmental benefits: "No doubt there are some conservation benefits to CRP lands—especially since it covers more than 36 million acres. However, that is like comparing a Walmart parking lot to a golf course. Just because it's green and has some plants, the golf course is better for wildlife than some pavement. Similarly, a plowed field planted [with] some row crop like corn is a biological desert...[so] taking any row crop out of production almost guarantees higher wildlife significance."[30] He also argued that conservation funds would be better spent on targeting critical wildlife habitats and buying those lands, which would yield long-term conservation benefits.

With grain prices relatively high in 2008, USDA came under pressure to allow producers to get out of their CRP contracts early without having to pay a penalty. This would have encouraged more farmers to withdraw their CRP lands and switch back to agricultural production. Secretary of Agriculture Ed Schafer announced on July 29 that he had decided not to allow CRP producers to withdraw from contracts early. A transcript of this announcement was posted on the USDA website, allowing people to see how the secretary of agriculture explained his actions.[31] The internet also allowed interest groups to post their reactions to this decision directly to the world. The Environmental Defense Fund website titled its posting "USDA Resists Pressure to Gut CRP" and explained that "conservationists and sportsmen's groups cheered a July 29 decision by USDA Secretary Ed Schafer to keep Conservation Reserve Program lands in conservation."[32] In the Sierra Club's "Wild Blog," Matt Kirby similarly marked this decision as good news. In his posting, he quoted Bart James of Ducks Unlimited as saying, "the Conservation Reserve Program is the holy grail of conservation, and we are pleased that the USDA will maintain the program and the benefits that it has had."[33]

The coexistence of the mainstream press, nonprofit information sources, and internet bloggers means that future coverage of the CRP will continue to be a conversation with more voices. Hard-news consumers will still find analyses, like the August 2008 *New York Times* article "As Prices Rise, Farmers Spurn Conservation Program." It pointed out that high crop prices meant that more farmers were deciding not to renew their CRP contracts once they expired, meaning that, in fall 2007, producers decided to take "back as many acres as are in Rhode Island and Delaware combined."[34] Pheasants Forever and Ducks Unlimited argued that USDA should raise CRP rates to keep land within the program. Those who use grain as inputs in their business, however, were quoted in the article as opposing any attempts to keep land within the CRP. As J.R. Paterakis, a baker from Baltimore, put it, "We're in a crisis here. Do we want to eat, or do we want to worry about the birds?" Jay Truitt of the National Cattlemen's Beef Association similarly asked, "Do you think it's right for you to pay so there's more quail in Kansas?"

The *New York Times* article was the beginning, rather than the end, of the conversation for some on the internet. A blogger named "Progressive Conservative" linked to the article and discussed his reaction to the piece, noting up front in his post that he was a member of Ducks Unlimited and Quail Unlimited and a hunter on restored habitat acres on a farm in Kentucky. He interspersed quotes from the article with his own assessments of the policy arguments. At the end of his analysis, he came to this conclusion: "I think, as with all issues, this one is going to require some compromise. What may be the best solution is to increase the dollar amount of rental payments to farmers, while slowly decreasing the amount of acreage being paid for. This will allow wildlife to adjust and help offset the loss of commodity revenue for farmers in the short run."[35]

If a website builds a conversation about farm policy on the internet, there is no guarantee that people will view it. Rational ignorance about the details of policy means that most people do not consume information about the CRP. Farmers, environmentalists, and sports and hunting advocates may seek out this information as part of their work or passion or hobby. Local newspapers provide information on the CRP's local implementation, and national newspapers cover the political battles about farm bill debates. The advent of the internet, however, means that those who want to know more can drill down to find out how USDA operates the program, how interest groups would change it, and how individuals experience it. The net also provides access to a very different type of information about the CRP—academic research on its implementation and effects.

ACADEMIC PERSPECTIVES ON THE CONSERVATION RESERVE PROGRAM

Although its aim has shifted from reducing soil erosion and stabilizing farm incomes to creating offsite environmental benefits and onsite wildlife habitats, the CRP has always operated at the macro level as a policy to change land use. One set of academic analyses looks at the CRP from this perspective and analyzes how the program has changed land use and what alternative policies might yield.[36] Studying land-use changes across six different categories of use (crops, pasture, forest, urban, range, and CRP) between 1982 and 1997, researchers found that decline in cropland area was driven primarily by the operation of the CRP and by drops in crop markets. The results, however, showed the importance of looking closely at how government payments affect producer incentives.

> By affecting the profitability of cropland, however, government [crop] payments may have had certain unintended environmental consequences, given that we identify pasture and CRP as the most important land-use margins for crops. In particular, we find that acreage in pasture and the CRP would have been almost 3.4 and 3.3 million acres greater in 1997 if government crop payments had been zero after 1978...This suggests that the government to an extent is directly competing with itself in providing incentives for landowners to retire environmentally sensitive cropland. In terms of the CRP, our findings illustrate the importance of measuring the impacts of a land-use policy, relative to a counterfactual baseline. We find that only about 90% of the lands that enrolled in CRP actually constituted "additional" land retirements induced by the policy, with the remaining 10% comprising lands that would have left crop production anyway, given the declining profitability of crop production.[37]

The availability of data on where CRP contracts have been struck between USDA and producers has given rise to detailed regional analyses of land use. Academics have modeled the impact of the CRP on land use in study areas ranging in size from the northeastern United States to the upper Mississippi River Basin to the state of Virginia.[38] The operation of the bidding process has given rise to empirical studies of potential problems with the program's operation. Researchers have studied whether there is a better mechanism to use in the CRP offer process and whether the existence of the CRP causes some farmers to bring additional land into crop use because of higher commodity prices or the substitution of noncropland into crop raising (a topic of opposing studies).[39] Some studies concentrate on easily recognizable impacts, such as the effect of the CRP on land values and rents.[40] Others have traced more indirect effects, such as the reduction in grain-elevator merchandising margins in Oklahoma brought about by the decline in grain supplies stemming from the operation of the CRP.[41]

A frequent type of CRP study examines producers as the decisionmakers and tries to determine what factors influence their decisions.[42] Surveys of CRP contract holders have allowed researchers to explore what plans these producers have for their lands after the conservation agreements expire and how rental rates might affect their decisions about renewing their CRP contracts. Others have studied how uncertainty and irreversibility in the setup of the CRP contracts affect the willingness of farmers to sign up. The emphasis on environmental justice in current policy research is also apparent in CRP research. In the article "Does Race Matter in Landowners' Participation in Conservation Incentive Programs?" the authors found some differences in participation in conservation programs: "Both white and minority landowners tended not to participate in conservation incentive programs and were equally likely to participate in the overall programs, Conservation Reserve Program (CRP), Stewardship Incentives Program (SIP), and Forestry Incentives Program (FIP). White landowners, however, were enrolled in the CRP longer and signed up more acres in the CRP and FIP than minorities. Moreover, minorities were more likely to be dissatisfied with program participation and to be unable to afford the cost share."[43]

As the selection of lands to incorporate into the CRP focused more on wildlife habitat benefits, articles on the impact of the CRP on birdlife began to appear. These were often written by wildlife agency researchers or professors in natural resource or environmental departments and focused on observations made about the effects in a particular state.[44] The studies most often compared the variety of species and density of populations on CRP versus non-CRP fields or the change in species variety over time in a given set of CRP fields. The results of these studies are nuanced and often linked to the particular type of cover on CRP lands. In Ohio, mowing CRP fields to control noxious weeds may interfere with nesting birds. In the prairie states, CRP lands are now home to species that had declined in previous decades. In central Iowa, CRP lands have replaced row-crop habitats that had lower bird densities and fewer nesting species. In Missouri, researchers found that among CRP fields those planted with shorter, more diverse cool-season grasses were better habitats for grassland birds than fields planted with tall, vertically-dense switchgrass. In Kansas, however, researchers found that CRP fields had more bare ground than pasture land and estimated that if CRP fields were transitioned to moderately grazed pasture, the impact would not be harmful to grassland birds.

Studies estimating the benefits of water quality improvements from the CRP tend to be done at the regional or national level and involve extrapolations from multiple sources of data.[45] To estimate the water quality benefits from reducing runoff from CRP lands, researchers have calculated the reduction in defensive expenditures and changes in production methods that result from cleaner water. If water contains less agriculture runoff, for example, water treatment costs may be lower and municipal and industrial water use costs can be reduced. Increased water quality may also increase use of lake and river recreation areas by people more likely to participate in activities, such as fishing and waterskiing, when waters contain less agricultural runoff.

Reduction in soil erosion, once the primary environmental benefit cited of the CRP, continues to attract attention from researchers.[46] An early study of the CRP signup process demonstrated that farmers in many regions did not appear to incorporate fully the productivity gains from reduced soil erosion when making decisions about enrolling in the CRP. Later studies explored how regulators could use ecosystem modeling and GIS technology to better target the lands taken out of production, if goals included reduced erosion and watershed protection. At the (literal) ground level, researchers explored how the composition of legumes and grass planted on CRP lands in Wyoming affected the recovery of soil organic matter.

As concern with global warming increased, the words "carbon" and "Conservation Reserve Program" began to appear in the same journal article titles, as people stressed the ability of forests to sequester carbon dioxide. A 1995 study estimated that a "program similar to the Conservation Reserve Program would sequester 48.6 million tons of carbon per year (3.5% of U.S. emissions) on 22.2 million acres."[47] That same year, researchers looking at various CRP scenarios concluded that "afforestation of marginal cropland... could provide approximately 15% of the C[arbon] offset needed to attain the Climate Change Action Plan of reducing greenhouse gas emissions to their 1990 level by the year 2000 within the United States."[48] Researchers noted that carbon sequestration effects from the CRP are potentially greater than the effects realized to date because this criterion has not played a large or long role in the debate over how lands should be targeted in the program.[49]

There is an extensive literature on the political economy of farm legislation with the theme that large campaign contributions translate into subsidies that benefit narrow organized interests at the expense of widely unorganized and dispersed taxpayers. Researchers have found that contributions are timed and managed to influence the course of legislation, noting that "the agricultural industry targeted contributions toward nonsenior, conservative Democrats from agriculturally dependent states who were expecting close election races and who served either on the agriculture committee or the agriculture appropriations subcommittee."[50] An overview of farm bill politics found that "empirical results revealed that rent seeking works, i.e., contributions influence agricultural subsidies in the manner they best serve contributors' economic interests. Eliminating campaign contributions would significantly decrease agricultural subsidies, hurt farm groups, benefit consumers and taxpayers, and increase social welfare by approximately $5.5 billion. Although contributions are not the only determinants of agricultural subsidies, investment returns to farm PAC [political action committee] contributors are quite high ($1 in contributions brings about $2,000 in policy transfers)."[51] Theoretical models of political influence note that, while environmental groups may not make as large contributions, they can wield political influence because of "their greater effectiveness in public persuasion and the growing public environmental awareness."[52]

There have only been a few studies that examine how information is produced and used within the CRP regulatory process. Authors from USDA and academia have combined to study the evolution of the EBI scoring system. One study describes the use of the EBI in the CRP in this way: "The program has a *monitoring* component, an index that *integrates* indicators, priority weights on each of the environmental objectives, which functions as an *assessment* tool, and a *management* plan for dealing with outcomes from the index."[53] They noted that, in the early CRP signups (e.g., through the 9th signup), there were criticisms voiced that the program focused too much on low productivity land in the Plains and on the onsite productivity losses from erosion. Once the 1990 Food Agriculture, Conservation, and Reform Act directed USDA to consider more environmental goals in enrollment, USDA developed the EBI.

In the early EBI, "each of the factors received an equal weight. Policymakers refused to explicitly judge the relative weights appropriate for each part of the index."[54] Starting with the 13th signup, however, "factors were weighted to reflect what were believed to be the most desirable outcomes of the program. The EBI was not meant to be a rigid index, but to be adjusted and improved depending on the progress of signups, perceived deficiencies, and/or changed priorities."[55] Analysis has shown that lands enrolled using the EBI system did score higher on this index than previously enrolled lands and that "most of this improvement is owed to improved wildlife habitat benefits and water quality benefits, and decreased rental costs due to enhanced competition for bids by landowners."[56]

The USDA and academic authors concluded that the EBI is flexible and allows for some attempts to target expenditures based on multiple dimensions of environmental benefits. They acknowledge, however, that there are no extensive studies that attempt to show how the EBI scores (which reflect a combination of science and valuation) translate into actual benefits that could be quantified and monetized.

A study by Resources for the Future (RFF) analyzed a risk assessment by USDA of the CRP as part of its compliance with the Federal Crop Insurance Reform and Department of Agriculture Reorganization Act of 1994.[57] The broad recommendation from these outside researchers was that USDA should try to conduct analyses aimed at estimating cost-effectiveness, rather than determining the set of natural resources at greatest risks. The authors noted that USDA faced difficulties in evaluating the CRP because Congress had not been explicit about all the environmental benefits to be valued and what weight to give to particular outcomes. The RFF researchers advised the agency to in the future seek information from focus groups and public opinion surveys as they brought in values and preferences in their analyses. Practical suggestions also included building up national estimates from regional models and including habitat diversity as a more prominent indicator of biodiversity.

The shifting, multiple priorities in the CRP and imperfections of available data in the program have generated some analyses which run scenarios to examine how USDA could operate the program differently. What might the results be if USDA tried to maximize enrolled acreage, maximize environmental benefits (or subcomponents relating to water erosion, wind erosion, groundwater vulnerability, or wildlife habitat), or maximize net benefits by taking into account the benefit-cost ratio in enrollment decisions?[58] These studies often find inefficiencies in the program and reach such conclusions as "implementation of the CRP in 1986 was suboptimal in the sense that net government cost of the program could have been reduced while simultaneously increasing the level of erosion reduction and supply control achieved."[59]

The most comprehensive accounting of the benefits and costs of the CRP probably is the final rule analysis in 1997, published in the *Federal Register*. Academics rarely attempt to add up the overall costs and benefits in one article, since their analyses are often focused on much smaller pieces of the puzzle (e.g., how are grassland birds in a particular state affected by conversion of land from row crops to CRP land). The legislative debates surrounding the details of the farm bill sometimes generate broad assessments. In the early days of the program (1989,) economists at USDA's Economic Research Service published (with coauthors) the article "CRP: What Economic Benefits?" They estimated that, for a 45-million acre enrollment, "the CRP will generate about $10 billion in natural resource benefits in present value terms. The largest share of the benefits is from improved wildlife habitat (40%), followed closely by surface water quality improvements (37%). Soil productivity, air quality, and groundwater supply benefits are less significant."[60] The authors were quick to point out that they did not attempt to estimate the net benefits (i.e., benefits minus costs), could not capture all categories of benefits and costs, and did not know if the benefits could be achieved at a lower cost through other policies. No recent academic study has attempted to estimate and monetize the net benefits of the current configuration of the CRP.

The Council for Agricultural Science and Technology adopted a different methodology for assessing the CRP during the 1995 debate over the program. The Council surveyed interest groups about the CRP, and 16 different representatives recommended six aspects of the program that could be modified: "enrollment size, targeting options, targeting tools, easement and contract usage, economic land use options, and delegation of control."[61] Ironically, these interest groups probably would have participated aggressively in the development of the EBI, had the formula been put up for debate in the *Federal Register*. Without that option, it was left to a nonprofit group—the Council—to survey associations about their general positions on the renewal of the CRP and desires for program changes.

CHAPTER 7

INFORMATION AND REGULATORY IMPLEMENTATION

Viewed from a distance, many regulatory programs look remarkably similar. A market failure arises, in part, because of lack of information. A government program is proposed, with Congress giving broad directions to a regulatory agency on how to remedy the market failure. The delegation of decisionmaking power to a set of regulators creates a new information problem, however. Members of Congress will need to worry whether executive branch officials in Washington, DC, and their agents in the field will really make the policy decisions the way that legislators want them to.

Worried about regulators' possible choices, legislators will set up a series of mechanisms to monitor what an agency is doing. This generates a flow of information back to Congress and lets department officials know their actions are being monitored. Actually knowing how the regulations are being interpreted on the ground is still difficult. Overall, this chain of delegation can be seen as a system of tradeoffs that involves accepting some slippage in policy decisions in return for lower costs of monitoring the decisions made by regulators. In this respect, regulations at their heart involve a substitution of government imperfections for market imperfections.

The U.S. Department of Agriculture's operation of the Conservation Reserve Program fits this description of regulation as trading one set of information problems for another. Markets will not lead farmers to incorporate fully into their decisions the negative results from cultivating their fields, such as soil and chemical runoffs that reduce water quality in nearby streams. These producers will not fully factor in the benefits from wildlife that flock to fields not planted with crops. Congress created versions of the CRP to remedy these problems by paying farmers to withdraw fields from cultivation and use them for conservation, but the marching orders in the legislative texts were usually amorphous. This meant that USDA had relatively wide discretion in designing the process of enrolling fields in the CRP. Many groups and institutions monitored this process, including GAO, congressional oversight committees, special interest groups, and reporters.

The process of notice-and-comment rulemaking usually offers Congress another way to monitor an agency's actions. The CRP differs from many other regulatory programs because one of its most central facets, the Environmental Benefits Index scoring system, did not undergo the standard notice-and-comment process. Yet, it appears as if USDA took the informal route of devising the EBI as a way to run a flexible, responsive, and evolving regulatory program. This chapter focuses on how the design of the CRP safeguards information generation and leads to the relatively successful delegation of decisionmaking power in this conservation program.

IMPERFECT INFORMATION

Driving down a rural road, car noise may startle ducks into action. As they fly away, the driver may notice the symmetry of their formation and appreciate the sunlight bouncing from their feathers. At home, a person drinks a glass of water, which seems to be free. But, in reality, the homeowner's water bill payment goes in part to a water treatment plant meant to make sure that what comes out the tap is potable. The water tastes great, although it may be easy to miss the trace elements of chemicals from pesticide and fertilizer runoff that has made their way into the water system.

This simple scenario illustrates the market failures surrounding croplands. Who do people thank when they see the birds fly across the sky, even though they know that the birds may find habitat on the land of a farmer who has chosen not to plant crops in a field? Who is to blame when the utility bill shows an extra charge for water treatment costs to deal with sediment and chemicals flowing into the water system? Whose fields generated the chemical runoff that shows up as those trace elements in the drinking water?

If information were free and negotiations were costless, these effects from a farmer's field and treatments for the crops planted might not be a problem. If the farmer had the property right to use chemicals and fertilizer on his field, nearby neighbors could costlessly (e.g., magically) know the damages caused by the sediments and pollutants flowing off the cropland. Each person could easily negotiate with the farmer and pay to reduce the harms experienced from the chemicals flowing from the fields. Neighbors might even write contracts that bound themselves to engage in these negotiations and payments, so that no one resident could free ride and let others shoulder the costs of paying the farmer not to pollute. The end result would be that the farmer would change the way he or she operated (by changing which land was planted with crops or changing how the fields were cultivated) until the marginal cost of reducing the amount of pollution coming from the farmer's field equaled the marginal benefit the neighbors received from reducing this stream of pollution from the fields.

In this same world, people who value knowing that ducks and pheasants and other wildlife can roam free on fallow ground could come together to pay the farmer to leave more fields in natural grasses and trees, rather than planting them with crops. The farmer would alter decisions about placing a parcel in crop production versus conservation based on the payments from those who cared about the wildlife on his fields. The negotiations would proceed until the marginal cost to the farmer of putting a unit of land into conservation (e.g., the net flow of funds the farmer would be giving up since he or she could not grow crops there) equaled the marginal benefit to him of this conservation (which would include payments from appreciative people who value the existence of wildlife habitats on his property).

This would truly be a magical world, where people took into account the spillovers of their actions on others. The world we live in, however, is very different. We do not know the exact harms arising from runoff from fields or the exact benefits of wildlife habitats. We cannot costlessly negotiate with agricultural producers about these benefits and costs, in part because we may not even know the identities of who is generating what positive and negative effects. We cannot rely on the hope that neighbors will voluntarily band together to contribute to pay producers to think about how their land decisions impact society. Information is costly. Transactions are costly. And people free ride, so one cannot simply assume they would contribute to changing farmers' decisions about cropland.

If someone were to design a set of institutions to deal with market failures and collective action problems, they would need to acknowledge the problems posed by the frailties of knowledge and human nature. Fortunately, the founders who designed the U.S. federal system of executive, legislative, and judicial branches did explicitly think about how the division of power and a system of checks and balances could build upon the way that people act in markets and politics. As James Madison noted in *Federalist No. 51* in 1788,

> What is government itself, but the greatest of all reflections on human nature? If men were angels, no government would be necessary. If angels were to govern men, neither external nor internal controls on government would be necessary. In framing a government which is to be administered by men over men, the great difficulty lies in this: you must first enable the government to control the governed; and in the next place oblige it to control itself. A dependence on the people is, no doubt, the primary control on the government; but experience has taught mankind the necessity of auxiliary precautions.[1]

In the instance of the CRP, solving the market failures posed by cropland spillovers involves creating a set of government institutions that build on Madison's concepts of delegated decisionmaking and widespread monitoring. Multiple principal-agent relationships are involved in the CRP. Voters delegate to members of Congress decisions about how to protect human health and the environment. Legislators write various versions of farm legislation which created and sustain the CRP. The legislation spells out the broad outline of how the program will work and often sets a limit on the total acres that can be enrolled. Producers with eligible lands are paid a rental fee to forego crop production on their land and manage these acres with conservation plans for 10–15 years. The objectives of the legislation vary over time: Congress first stressed the maintenance of farm incomes, management of commodity supplies, and reduction in soil erosion as the benefits of the CRP. During the 1990s, the emphasis shifted to enrolling lands that could generate other environmental benefits, such as improved water quality and increased wildlife habitats.

The inclusion of the CRP in farm legislation helps members of Congress craft a winning legislative coalition because it attracts support from both environmentalists and hunters. Congress does not spend the time or legislative capital to hammer out the exact details of how the CRP will operate. It has effectively delegated this to USDA. At USDA, the Farm Service Agency and the Commodity Credit Corporation (the body that contracts with farmers for the land) basically have been given wide discretion by Congress. These regulators develop the expertise to figure out how to design a bidding system for lands and how to design a scoring metric to measure a parcel's potential environmental benefits. Legislators are free to devote their attention to other aspects of farm legislation, and FSA officials can focus on program design.

While delegation creates obvious benefits for legislators, the operation of the principal-agent relationship also carries costs. Members of Congress must take into account that USDA officials may not make the "right calls" in setting up and operating the CRP. Legislators could examine and second-guess every decision down to the rental contract level, but such micromanagement obviates any advantages in delegation. If the legislator is always looking over the regulator's shoulder, there are no gains from giving the regulator the responsibility to make the decision. The freedom of the regulator to act, however, carries two costs. The possibility of hidden action means that Congress may not be able to see every action that FSA agents take to implement the CRP. If a fraction of these decisions go against Congress's intentions or legislators' constituents, then that is a cost. There are also the dangers posed by hidden information. Members of Congress may not be able to know what information FSA officials had, when they adopted a bidding process or created a scaling standard.

The scale of the CRP is immense. In 2008 it covered more than 33 million acres, in more than 700,000 contracts, struck with more than 400,000 farmers.

In 2008 more than $1.7 billion in rental payments flowed to producers participating in the CRP. The challenge for Congress in overseeing USDA's operation of the CRP lies in trading the costs of monitoring agency action with the costs of agency discretion. If Congress ignored the CRP, then FSA could, in theory, operate the program to benefit a narrow set of stakeholders (e.g., farmers in particular regions of the country or those who valued specific types of wildlife habitats). If members of Congress and their staffs constantly reviewed CRP decisions, there would be little agency discretion and few gains from delegation. A review of the history of the CRP shows, however, that there are numerous ways that Congress can learn about the operation of the CRP without investing extensive amounts of its own staff time and resources in the oversight.

MONITORING AND DISCRETION

One way that Congress economizes on providing policy directions is to delegate decisions to an agency and then review its performance when a policy's authorizing legislation is up for revision. The omnibus farm legislation that authorizes the CRP is renewed every five to six years. To craft new farm legislation, the House and Senate agricultural oversight committees and subcommittees often hold hearings on the CRP. At times, these have been held in a legislator's home state or district. This gives regulatory oversight a tinge of constituency service because it invites local stakeholders, interest groups, and constituents to present their views and interact with agency officials about the implementation of the program. From the perspective of U.S. Senators and Representatives, these oversight hearings entail a cost in organizing and an opportunity cost of time for participation. A large segment of the information exchanged, however, is created and communicated by interest groups (e.g., those representing environmentalists, hunters, farmers, and agribusiness operators).

The congressional oversight hearings allow legislators to send signals to the department and its agencies. The questions they ask reveal areas where members of Congress may be displeased by CRP policies. Frequent topics of discussion in the CRP hearings, for example, include whether appropriate attention is placed on soil erosion, wildlife habitat, water quality, and the regional distribution of conservation payments. These signals are also helpful to regulators because hearings may be scheduled between the announcement of a proposed or interim rule and the issuance of a final rule. Although the Environmental Benefits Index was not issued through a formal notice-and-comment rulemaking, congressional hearings did provide USDA the opportunity to submit a description of the EBI's calculation as testimony and allowed members of Congress to question officials about how the EBI would work in the field.

Congress at times delegated investigation of the CRP's operation to the General Accountability Office. House and Senate oversight committee chairs often asked GAO to investigate the CRP's operation and cost-effectiveness. GAO gathered information by analyzing results of interest group letters to Congress about the CRP, surveying state committees involved with CRP implementation, and conducting audits of farm program payments. From USDA's perspective, the GAO reports were like police patrol monitoring. GAO investigators and analysts would examine individual subsets of the agency's decisions and report to Congress on the outcomes.

USDA's Office of the Inspector General used both police patrol and firm alarm monitoring in its examination of the Conservation Reserve Program. The Inspector General conducted regional audits, where investigators looked closely at the terms of conservation rental agreements and determined whether the enrolled lands actually qualified for admission into the CRP. One advantage of these audits for the Farm Service Agency is that they were internally circulated, rather than widely cited like the usual GAO reports. The Office of the Inspector General's hotline provided an additional source of information to USDA officials in Washington on how the CRP was working in the field. The calls to the hotline, at times anonymous, caused the Inspector General to investigate claims that, in particular regions or offices, the regulators were incorrectly interpreting regulations or missing instances of fraud by CRP producers.

Passage of the omnibus farm legislation would often be followed by a proposed or interim rule governing the Conservation Reserve Program. USDA placed two of the most important questions about the CRP—the design of the bidding process mechanism and the scoring process that evolved into the Environmental Benefits Index—outside the notice-and-comment rulemaking process involved with publication in the *Federal Register*. By publishing the broad outline of the CRP policies in the *Federal Register* and seeking comments, however, USDA accomplished several goals. It provided many interest groups and stakeholders the opportunity to express support for or frustration with aspects of the CRP. When a final rule on the CRP was published in the *Federal Register* in 1997, for example, the agency was able to respond to over 3,400 comments on the proposed rule. The rulemaking process served as a safety valve for stakeholders to let FSA know about problems and as a source for freely provided ideas and information for the agency (and congressional principals that also monitored the *Federal Register* and read about the progress of the rulemaking).

In academic theories of the regulatory process, principals often rely on third parties to report on the activities of their agents. In the implementation of the CRP, the Environmental Working Group provided detailed information on USDA's behavior to two sets of principals— Congress and voters. Using the Freedom of Information Act and a good set of negotiating skills, EWG liberated detailed data from USDA on exactly who received payments under the commodity subsidy and conservation payment programs. EWG then analyzed the data in a way that reframed congressional debate about agricultural policy. EWG's reports showed the regional distribution of CRP funds, the high concentration of subsidy payments in commodity programs in particular districts, the large fraction of USDA funds which ended up in the top 10% of producers receiving aid, and the relatively low level of payments in programs, such as the CRP, that went to family farms. USDA could have performed the exact same analysis as EWG since the data came from the department, but EWG was more willing to provoke political controversy and publish analyses that pushed for reform.

From the perspective of members of Congress interested in the performance of FSA, the work of EWG was a gift. Private foundations that supported the data-intensive work of EWG and believed in the power of information to influence policy debate sustained these analyses. Congressional staff, along with reporters, interest group members, and interested rural neighbors, could search the EWG database without having to contribute to its creation or maintenance. The public and legislators also benefited from the ability of EWG to evolve as technology changed. This environmental group went from producing hard-copy reports and press releases based on analysis of mainframe computer tapes to posting on the internet an array of information that included a hugely successful, searchable database on farm payments, a policy blog about the path of farm policy, and a set of videos on *YouTube.com* about the group's agricultural (and other environmental) interests.

Media reports are another source of monitoring information for Congress members concerned about the delegation of discretion to agencies. The most distinctive aspect of media coverage of the CRP lies in the tendency of local media to focus on the local aspects of policy implementation. EWG courted coverage of its analyses by calculating the effects of farm policy payments in states and congressional districts. This lowered the cost for regional and local newspapers in creating a story and increased the likelihood that the reporters would write about the CRP. Hunting groups also developed information on the wildlife habitats on local CRP lands to generate news stories. Politicians seeking local credit with their constituents were happy to provide quotations to local reporters writing about the progress of successive farm bills and inherent prospects for the CRP. The information in local newspaper articles also served the producer and consumer interests of readers. Local stories about the CRP at times advised farmers on where and when to sign up for the program and local hunters on where the CRP had improved local wildlife habitats.

As a multibillion dollar program affecting hundreds of thousands of people, the CRP attracts its share of monitoring. The oversight may take the form of congressional hearings, GAO reports, audits by USDA's Inspector General, rulemaking comments, interest group reports, and journalists' articles. What is striking about the operation of the CRP, however, is how USDA has maintained such discretion in the operation of the program that it can change the implementation of the policy as new priorities emerge, technologies change, and regulators learn from feedback in Washington and the field.

Central to USDA's flexibility in administering the CRP is its decision to keep the development and deployment of the EBI outside the formal rulemaking process. For parcels offered by producers during the general CRP signup, it is the detailed metrics in the EBI that determine whether the lands make it into the CRP. Congress only provided USDA with very broad criteria to use in selecting lands, for example, directions to consider soil erosion, water quality, and wildlife habitat benefits on proposed fields. Even though interest groups (and even legislators) complained that the particular weighting systems disadvantaged local landowners in their states and districts, fair-minded legislators knew that it was the broad discretion that they gave the agency that made the scoring process possible.

For USDA, publishing the EBI in agency guidance documents that are available at local FSA offices, rather than publishing it for comment in the *Federal Register* made perfect sense. Informal publication meant that the agency could change the EBI easily between signups, such as recalculating the weights of particular environmental benefits as priorities changed. The agency also changed the metrics and questions asked when officials saw that some questions in particular generated human errors in calculation or interpretation. The development of centralized databases, GIS technology, and easily accessible satellite imagery now means that producers can quickly and transparently see how their EBI scores are derived. USDA officials hope that this allows farmers to maximize the environmental benefits of their enrollment proposals and minimizes the costs to the government of the bid payments in the CRP rental agreements.

The freedom of USDA to keep the EBI out of the *Federal Register* and the notice-and-comment rulemaking process may stem from the fact that agency officials and most stakeholders view the program as a voluntary process. USDA officials are quick to point out that farmers are not forced to participate in the CRP. They note that when the EPA regulates a plant, the EPA will often use the force of law to require a facility to change its operation, also known as "command and control" regulation. In the CRP, however, agency officials essentially write the rules that govern how the government distributes gains to producers who choose to put their lands in conservation. The voluntary nature of the program, and the fact that it distributes gains rather than imposing losses, gives USDA the freedom to keep its key regulatory decisions flexible. The bidding process to enter the CRP and the types of factors valued by the agency are now fully transparent to stakeholders and potential participants. The agency's ability to keep these policy decisions out of formal rulemaking ensures that it can keep modifying the EBI with changes in priorities, information, and technology. The provisions of the Administrative Procedure Act that exempt contracting details from formal rulemaking provide a legal justification for keeping the changes made to the EBI out of the notice-and-comment process.

How well has the CRP worked? The continued willingness of producers to enroll their parcels in this conservation program shows that from their perspective the program offers benefits. From society's perspective, the question arises about whether the CRP changes behavior and increases environmental benefits. Academic studies indicate that the CRP drew cropland into conservation, which at times increased commodity prices and farm incomes and created more habitat for some wildlife. The rough sorting of costs and benefits required in the rulemaking process credits the CRP with ultimately generating net benefits for society. Much of the academic literature and some government analyses focused on a different question of whether more environmental benefits could be delivered for the same level of costs. As more data on environmental outcomes, satellite images of land use and wildlife, and algorithms for estimating environmental benefits become available, there are prospects for even better targeting in the program. This might arrive, for example, through changes in the EBI and changes in technology at the local office level that would provide greater fine tuning of CRP offers.

From the perspective of mechanism design, the CRP works remarkably well. The dispersed but detailed monitoring that goes on in the program means that Congress has successfully conserved a key resource in the design of this policy—information. As people drive down a rural byway, they are rationally ignorant about the details of farm policy and the CRP. They may appreciate the scenery and the nearby wildlife, but be completely unaware of the details of rental contracts and EBI scores. The members of Congress who make policy choices for voters also are free to operate without devoting much daily attention to the CRP. They set the broad direction of the CRP and limits on eligible acres during the crafting of farm legislation, and then delegate discretion to USDA to implement the program. If Congress members wanted complete control over the implementation of the CRP, they could order costly investigations and audits of nearly every decision. Instead, legislators are willing to conserve on information and rely on the structure of institutions. The transparent regulatory process and widespread monitoring by stakeholders combine to yield conservation policies that continue to win favor among a wide set of observers—farmers, environmentalists, hunters, members of Congress, and agency officials.

LESSONS LEARNED

Throughout the policymaking cycle involved in the development and implementation of the CRP, legislators and regulators make decisions with imperfect information. From the perspective of an individual, such as North Carolina farmer James Grass whose CRP lands are described in the introduction, there may be frustration with the lack of perfect information. Why does the government not know more about the exact types of benefits and costs coming from different parcels of land and use this information to determine the optimal mix of conservation plans? From the broader perspective of the program as a whole, however, the tradeoff between precision and information costs becomes clearer. More resources could be devoted to measuring and monitoring in the CRP, which could lead to better targeting of conservation efforts. But, this knowledge may come at the cost of less funding to rent the land from farmers for conservation.

Stories about the development of a regulation are not often told. Academic analyses are more likely to focus, for example, on the specification of cost-effectiveness or distributional impacts of a policy. This book tries to examine the evolution of a rule by showing how decisions were made and how they changed over time, given the information that developed about the implementation of the CRP. This flow of information is part of all regulatory tales. A single case study of the CRP cannot prove the general impact of information on agency decisionmaking. The particular story of the CRP, however, offers hypotheses about the role of data in policymaking that can be explored in other regulatory settings.

Six lessons about the role of information in regulatory policy arise from this analysis of the CRP.

Lesson 1. Regulation often involves substituting imperfect information about government decisions for imperfect information about market decisions.

Problems with knowing, defining, and allocating the property rights to the positive and negative spillovers from farmlands (new wildlife habitat or streams fouled by pesticide runoff, respectively), make it hard for the market to yield the optimal levels of conservation. The CRP helps solve these problems, but the implementation of such a large regulatory program also creates its own set of problems with imperfect information.

Lesson 2. The delegation of decisionmaking to regulators by Congress means there is a tradeoff between spending resources on program goals versus spending resources to monitor the implementation of the program.

By authorizing the CRP through very broad legislative language, Congress can delegate decisions to agency officials about what types of land to conserve. In theory, legislators can demand detailed information on exact conservation expenditures in the CRP and direct agency officials on what lands to buy. The resources devoted to monitoring what is going on in the field, however, reduce the resources for conserving more land. Legislators thus face a choice between the degree of full information they receive, which is costly, and the degree they are willing to delegate decisions and rely on imperfect data to monitor agency decisions.

Lesson 3. By making their goals and decisionmaking criteria more transparent, regulators can increase the likelihood that people will understand and comply with agency program requirements.

Once FSA released detailed information on how the EBI scored lands for admission into the CRP, farmers were better able to target lands proposed for conservation. The ability of farmers to see how different tracts and conservation practices led to higher EBI scores fostered competition to create (or preserve) the benefits explicitly valued in the EBI indicators.

Lesson 4. Regulators learn over time.

The flow of information generated by congressional hearings, press coverage, interest group reports, rulemakings, and academic studies provides agency officials with feedback on policy implementation. Changes in the operation of the CRP, such as alterations to the EBI, show how FSA officials can act on new information without waiting for explicit signals from Congress about the desirability of altering how policy is implemented.

Lesson 5. Regulators in contracting programs, in which people choose to participate, may often have more flexibility than those operating in command-and-control regulatory programs.

The ability of FSA to revise the EBI frequently allows it to change the emphasis on particular types of environmental benefits as it determines what lands to conserve through CRP contracts. Because the contracting process is voluntary—farmers choose whether to submit a proposal to put lands in conservation—FSA does not experience the extensive rulemaking battles that often occur with regulatory programs where an agency imposes requirements on individuals or firms.

Lesson 6. Interest groups can extract data from regulators and use this information to focus attention on issues that regulators may be less willing to emphasize.

Officials implementing the CRP do not have strong incentives to publicize the individual contracts in the program. Data on who gets how much from government programs raises issues of equity for politicians (e.g., why does a neighboring congressional district get more funds?) and privacy for farmers (e.g., why should neighbors know the value of each others' government checks?). Through the Freedom of Information Act, however, the Environmental Working Group gained access to agricultural subsidy data, including data on the CRP, and made it widely accessible via the internet. This in turn focused attention on allocation of funds within the CRP.

While these lessons help explain the general evolution of the CRP, they also help explain the particular experience of James Grass with the program. Grass was frustrated that improper site preparation meant that grass choked out the hardwood seedlings planted on his CRP lands. He was also disappointed that the Farm Service Agency appeared to continue to list the hardwoods as healthy trees on his land. He believed that some pine trees died on his lands because government contractors used machine planting rather than hand planting. In the future, innovations in satellite imagery and agricultural databases may make it more likely that government data and plans for CRP lands will prevent the problems Grass had with the program. The parcels brought into conservation and plans used to conserve such land will be based on information of increasing quality and quantity. This in turn will allow more tailored conservation approaches that better match the specific characteristics of each parcel within the CRP.

Current Farm Service Agency data that describe what is growing on Grass's farm may be imperfect. The planting plan for a set of the trees on his land was not ideal. Despite these glitches in data and plantings, overall the lands that Grass placed in the CRP look very much like those envisioned by legislators and regulators in Washington—rows of newly grown trees that are home to a range of wildlife. Even without full information about what happened, if legislators who voted for the CRP and FSA officials who helped develop the Environmental Benefits Index showed up at this North Carolina farm, they would appreciate and approve of the soil and wildlife conservation benefits there that are due to the Conservation Reserve Program.

NOTES

INTRODUCTION

1. To protect the privacy of the North Carolina producer interviewed for this book about his participation in the Conservation Reserve Program, I have changed identifying details about the farmer and his land. Rather than use his real name, I refer to him by a pseudonym, James Grass.
2. USDA, Farm Service Agency (2008).
3. Effective July 7, 2004, the U.S. General Accounting Office changed its name to the U.S. Government Accountability Office. (Both have the same acronym, GAO.) The current name is used when spelled out in this book.

CHAPTER 1

1. The discussion of principal-agent relationships here draws heavily on this work.
2. Hamilton (2004, ch. 6).
3. *Opensecrets.org*. Found in "Big Picture," "Price of Admission;" choose from menus: "Members running for re-election," Election cycle: 2006, "Display averages," http://www.opensecrets.org/bigpicture/stats.php?cycle=2006&Type=R&Display=A
4. *MAPLight.org* (2008). Found in "Press Room," "Reports," "Remote Control," http://www.maplight.info/remotecontrol08/RemoteControl08Report.pdf
5. Brehm and Hamilton (1996).
6. *Opensecrets.org*. Found in "Influence and Lobbying," "Industries," "Background: Agribusiness," http://www.opensecrets.org/industries/background.php?cycle=2008&ind=A
7. USDA, Farm Service Agency (2008).
8. Kahneman et al. (1991).
9. See http://www.archives.gov/federal-register/laws/administrative-procedure/553.html
10. USDA, Farm Service Agency (2008).

CHAPTER 2

1. Unless otherwise noted, all quotations in this chapter come from interviews with agency officials conducted for this book in fall 2007.
2. Dicks et al. (1987, *1*).
3. February 11, 1987.
4. Ibid.
5. Dicks et al. (1988, *63*).
6. Food Agriculture Conservation and Trade Act of 1990, Part 4 of 11 (1990, *3581*).
7. Barrett (1990).
8. *Congressional Quarterly* (1990) and Cloud (1990).
9. U.S. House of Representatives (1997a, 77–88). I am indebted to Ben Hendricks for analyzing the information presented in the hearing and devising this way to summarize the composition of the Environmental Benefits Index. The summary draws heavily on the language of the hearing appendix regarding how the EBI is calculated.

10. The Coastal Zone Management Act of 1972 (Public Law. 92-583, 86 Stat. 1280, enacted October 27, 1972, 16 U.S.C. §1451–1464, Chapter 33.
11. The 1972 amendments (Clean Water Act) to the Federal Water Pollution Control Act. Note that this includes plans created in accordance with State 305 (b) reports, State 303 (d) reports, 319 priority water lists, and plans green-lighted under section 320.
12. Zinn (2003).
13. USDA, Economic Research Service (1997).
14. Ibid. Note that this description of changes in the EBI includes direct quotations of terms from the article and paraphrases of the changes in EBI scoring.
15. USDA, Farm Service Agency (2008).
16. USDA, Farm Service Agency (1997, *7601–35*).

CHAPTER 3

1. USDA, Farm Service Agency (2008).
2. USDA, Office of the Inspector General (1990a, *1–2*). Farm Service Agency county committees consist of elected farmers and ranchers who provide local input on USDA programs, including conservation programs.
3. USDA, Office of the Inspector General (1990b, cover memo, *1*).
4. USDA, Office of the Inspector General (1996b, cover page)
5. USDA, Office of the Inspector General (1996a, *i*).
6. USDA, Office of the Inspector General (1996b, *i–ii*).
7. USDA, Office of the Inspector General (1995, *i*).
8. USDA, Office of the Inspector General (1998a, *iii*).
9. USDA, Office of the Inspector General (1998b, *7*).
10. USDA, Office of the Inspector General (2000a, *6*).
11. USDA, Office of the Inspector General (2008, memo, November 2, 2007, *1*).
12. Calls to the USDA Office of the Inspector General hotline were transcribed by USDA and may contain some minor, unintentional variances from the actual call.
13. USDA, Office of the Inspector General (1998c, *3*).
14. USDA, Office of the Inspector General (2000b, memo, September 27, 2000, *1*).
15. USDA, Office of the Inspector General (2007, memo, March 5, 2007, *1*),
16. USDA, Office of the Inspector General (2002).
17. USDA, Office of the Inspector General (2006).

CHAPTER 4

1. GAO (1989, *4*).
2. GAO (1992, *12*).
3. GAO (1995b, *5*).
4. GAO (1995a, *3*).
5. GAO (1999, *2*).
6. GAO (1998, *15*).
7. GAO (2002).
8. GAO (2003, *7*).
9. GAO (2004, *5*).
10. GAO (2005, *15*).
11. U.S. Senate (1988, *2*).
12. Ibid., *23*.
13. Ibid., *29*.
14. U S. House of Representatives and U.S. Senate (1994, *6*).
15. Ibid., *25*.
16. U.S. House of Representatives (1996, *2*).
17. Ibid., *24*.
18. Ibid., *4*.
19. Ibid., *30*.
20. Ibid., *40*.
21. Ibid., *42*.
22. U. S. House of Representatives (1997a, *2*).
23. Ibid., *20*.
24. Ibid., *35*.
25. Ibid.
26. U.S. House of Representatives (1997b, *19*).
27. U S. House of Representatives (1999) and U.S. House of Representatives (2000).
28. U.S. Senate (2006, *9*).
29. Ibid., *36*.
30. USDA, Agricultural Stabilization and Conservation Service, Commodity Credit Corporation (1986, *8780*).
31. USDA, Agricultural Stabilization and Conservation Service, Commodity Credit Corporation (1987).

32. Ibid.
33. USDA, FSA (1997).
34. Ibid.
35. USDA, Commodity Credit Corporation (2003).
36. Ibid.
37. USDA, Commodity Credit Corporation (2004).
38. Ibid.

CHAPTER 5

1. USDA, Farm Service Agency (2008).
2. Environmental Working Group (2008d).
3. Cook (2004).
4. Environmental Working Group (2008e).
5. As cited in U.S. House of Representatives (1994, *99*).
6. Ibid., *100*.
7. Ibid.
8. U.S. House of Representatives (1994, *43*).
9. Ibid., *48*.
10. Ibid., *46–7*.
11. Ibid., *99*.
12. Lambrecht (1994).
13. Hendee (1994).
14. Ibid.
15. *U.S. Newswire* (1997).
16. Ibid.
17. Ibid.
18. Ibid.
19. Ibid.
20. Ibid.
21. Environmental Working Group (2005).
22. Goodbody (2005).
23. Cook (2007a).
24. Cook (2008).

25. Ibid.
26. Cook (2007a).
27. Barrionuevo (2007).
28. Lochhead (2007).
29. Ibid.
30. Environmental Working Group (2008b).
31. Environmental Working Group (2008f).
32. Environmental Working Group (2008g). Quotes from numerous media reports about the EWG's operation as cited on the "About EWG" web page.

CHAPTER 6

1. Associated Press (1987).
2. Robbins (1987).
3. McAllister (1988).
4. *St. Louis Post-Dispatch* (1989a).
5. *St. Louis Post-Dispatch* (1989b).
6. Anderson (1994).
7. Anderson and Schara (1994).
8. Porter (1995).
9. Schara and Anderson (1994).
10. Goodsell (1995).
11. Beeder (1994).
12. Gordon (1995).
13. Marshall (1996).
14. Ibid.
15. Cowles (1997).
16. D. Smith (1997).
17. Porter (1997).
18. *New York Times* (2000).
19. *Rocky Mountain News* (2002).
20. White House (2004).
21. Porter (2003).

22. D. Smith (2004).
23. Ibid.
24. D. Smith (2004) and Mitchell (2008), respectively.
25. Herszenhorn (2007).
26. *Bismarck Tribune* (2008).
27. *Congressional Quarterly* (2008a; 2008b) and Richert (2008).
28. *Birding Business News* (2008).
29. *Jesse's Hunting and Outdoors.com* (2008).
30. Wuerthner (2008).
31. USDA, Office of Communications (2008).
32. Environmental Defense Fund (2008).
33. Kirby (2008).
34. Streitfeld (2008).
35. *ProgressiveConservative.com* (2008).
36. Alig et al. (1998), Feng et al. (2003), Osborn (1993), Olson (2001), and Parkhurst and Shogren (2003).
37. Lubowski et al. (2003, *29*).
38. Parks and Schorr (1997), Feng et al. (2005), and Siegel and Johnson (1991).
39. Roberts and Bucholtz (2005), R. Smith (1995), and Ervin and Dicks (1988).
40. Shoemaker (1989).
41. Adam et al. (2004).
42. Cooper and Osborn (1998), Brorsen (1996), Isik and Yang (2004), and Ibendahl (2004).
43. Gan et al. (2005, *431*).
44. Klute et al. (1997), McCoy et al. (2001), Swanson et al. (1999), Johnson and Schwartz (1993), Patterson and Best (1996), Nomsen (2005), and Dunn et al. (1993).
45. Ribaudo (1989), Feather and Hellerstein (1997), and Ribaudo (2004).
46. Miranda (1992), Yang et al. (2005), Lant et al. (2005), and Robles and Burke (1997).
47. Parks and Hardie (1995).
48. Barker et al. (1995).
49. Feng (2004).
50. Van Doren et al. (1999). See also Stratmann (1992a; 1992b; 1995; 1998).

51. Lopez (2001).
52. Yu (2005).
53. Ribaudo et al. (2001, *12*).
54. Ibid., *15*.
55. Ibid.
56. Ibid., *17*.
57. Powell and Wilson (1997).
58. Babcock et al. (1996; 1997).
59. Reichelderfer and Boggess (1988, *9*).
60. Ribaudo et al. (2004, *246*).
61. Hughes et al. (1995, *1*).

CHAPTER 7

1. Madison (1788).

REFERENCES

Abler, David G. 1991. Campaign Contributions and House Voting on Sugar and Dairy Legislation. *American Journal of Agricultural Economics* 73(1): 11–7.

Adam, Brian D., Seung Jee Hong, and Michael R. Dicks. 2004. Effects of the Conservation Reserve Program on Elevator Merchandising Margins in Oklahoma. *Journal of Agricultural and Applied Economics* 36(1): 83–96.

Alig, Ralph J., Darius M. Adams, Bruce A. McCarl. 1998. Impacts of Incorporating Land Exchanges between Forestry and Agriculture in Sector Models. *Journal of Agricultural and Applied Economics* 30(2): 389–401.

Anderson, Dennis. 1994. CRP Isn't Just for Farmers; Renewal of Program to Benefit Wildlife, Environment. *Star Tribune* (Minneapolis, MN), March 4, 4C.

Anderson, Dennis, and Ron Schara. 1994. Thanks to CRP the Good Old Days Are Back; 1994 Pheasant Opener. *Star Tribune* (Minneapolis, MN), October 9, 20C.

Arnold, Douglas. 1992. *The Logic of Congressional Action*. New Haven, CT, USA: Yale University Press.

Associated Press. 1987. Farmers Respond to USDA Offer. *Journal of Commerce*, February 20, 12A.

Babcock, Bruce A., P.G. Lakshminarayan, Jun Jie Wu, and David Zilberman. 1996. The Economics of a Public Fund for Environmental Amenities: A Study of CRP Contracts. *American Journal of Agricultural Economics* 78(4): 961–71.

_______. 1997. Targeting Tools for the Purchase of Environmental Amenities. *Land Economics* 73(3): 325–39.

Barker, Jerry R., Greg A. Baumgardner, David P. Turner, and Jeffrey J. Lee. 1995. Potential Carbon Benefits of the Conservation Reserve Program in the United States. *Journal of Biogeography* 22(4/5): 743–51.

Barrett, Amy. 1990. The Greening of the Congressional Consciousness Is an Indisputable Fact. *Roll Call* Policy Briefing, no. 13. Washington, DC: Roll Call.

Barrionuevo, Alexei. 2007. Agriculture Dept. Urges Big Overhaul in Farm Policy. *New York Times*, February 1, A20.

Batie, Sandra S., and Richard D. Horan, eds. 2004. *Economics of Agri-environmental Policy*. Vol. 2. Burlington, VT, USA: Ashgate.

Beeder, David C. 1994. Farm Bureau Hails Extension of CRP. *Omaha World Herald*, December 17, 51.

Birding Business News. 2008. Farm Bill News from the Bird Community Bulletin: Eventual Farm Bill Resolution. June 10. http://birdingbusiness.blogspot.com/2008/06/farm-bill-news-from-bird-community-e.html. Accessed September 2009.

Bismarck Tribune. 2008. Opinion. *Bismarck Tribune*, January 6, 7C.

Brehm, John, and James T. Hamilton. 1996. Noncompliance in Environmental Reporting: Are Violators Ignorant, or Evasive, of the Law? *American Journal of Political Science* 40(2): 444–77.

Brorsen, B. Wade, Carl Garrison, Brian D. Adam, and Michael R. Dicks. 1996. Nonresponse Bias Corrections for the 1990 SWCS Survey of Conservation Reserve Program Contract Holders. *Review of Agricultural Economics* 18(4): 669–80.

Coase, Ronald. 1960. The Problem of Social Cost. *Journal of Law and Economics* 3: 1–44.

Cloud, David S. 1990. House to Environmentalists: No more. *Congressional Quarterly*, September 29, 3112.

Congressional Quarterly. 1990. Congress Enacts Lean Farm Package. *Congressional Quarterly Almanac* 46 (1990): 334–5.

________. 2008a. Farm Bill Agreement. Congressional Quarterly House Action Report, no. 110-7, May 4, 13.

________. 2008b. Third Supplement to the Legislative Week of June 16, 2008. Congressional Quarterly House Action Report, June 18, 2.

Cook, Kenneth A. 2004. Farm Policy for the Rest of Us. Speech to the American Bankers Association Agricultural Bankers Annual Meeting, Minneapolis, MN, November 16. Available from http://www.ewg.org/node/8695

_______. 2007a. What's in This Database? In *Mulch*, posted June 12. http://www.mulchblog.com/2007/06/whats-in-this-database.html.

_______. 2007b. New Farm Bill Web Site Generates Media Storm. In *Mulch*, posted June 12. http://www.mulchblog.com/2007/06/new-farm-bill-web-site-generates-media-storm.html.

_______. 2008. Farm Bill: Any Means (Testing) to an End. In *Mulch*, posted May 6. http://www.mulchblog.com/2008/05/farm-bill-any-means-testing-to-an-end.html.

Cook, Ken, and Chris Campbell. 2008. Amidst Record 2007 Crop Prices and Farm Income Washington Delivers $5 Billion in Subsidies. *Environmental Working Group* website. http://farm.ewg.org/farm/dp_text.php. Accessed September 2009.

Cooper, Joseph C., and C. Tim Osborn. 1998. The Effect of Rental Rates on the Extension of Conservation Reserve Program Contracts. *American Journal of Agricultural Economics* 80(1): 184–94.

Cowles, Ann. 1997. Deadline Nears for Farmers to Reap Benefits of Program. *Atlanta Journal-Constitution*, March 20, 4I.

Dicks, M.R, K.H. Reichelderfer, and W.G. Boggess. 1987. "Implementing the Conservation Reserve Program." Washington, DC: USDA.

Dicks, Michael R., Felix Llacuna, and Michael Linsenbigler. 1988. The Conservation Reserve Program: Implementation and Accomplishments, 1986–87. Washington, DC: USDA, Economic Research Service.

Downs, Anthony. 1957. *An Economic Theory of Democracy*. New York: Harper.

Dunn, Christopher P., Forest Stearns, Glenn R. Guntenspergen, and David M. Sharpe. 1993. Ecological Benefits of the Conservation Reserve Program. Conservation Biology 7(1): 132–9.

Environmental Defense Fund. 2001. Losing Ground: A State-by-State Analysis of America's Growing Conservation Backlog. http://www.environmentaldefense.org/documents/702_losing%20ground.htm. Accessed September 2009.

_______. 2008. USDA Resists Pressure to Gut CRP. http://www.edf.org/article.cfm?contentID=8204. Accessed September 2009.

Environmental Working Group. 2005. New EWG Farm Subsidy Database Reignites Reform Efforts. November 1. Photocopy available from author. http://www.ewg.org/. Accessed 5 January 2006.

_______. 2008a. As Congress Finalizes Farm Bill Deal EWG Lists Recipients of Controversial Direct Payment Subsidies for 2007, April 29. http://farm.ewg.org/farm/dp_release.php. Accessed October 2009.

_______. 2008b. Congress poised to cut conservation funds that aided Farm Bill's passage, September 8. Available from http://www.ewg.org/book/export/html/27103. Accessed October 2009.

_______. 2008c. Farm Subsidy Database: Conservation Reserve Program in United States. Available from http://farm.ewg.org/farm/progdetail.php?fips=00000&progcode=total_cr. Accessed October 2009.

_______. 2008d. EWG Farm Subsidy Database Update, April 14. http://farm.ewg.org/farm/summary.php. Accessed October 2009.

_______. 2008e. How EWG Does It. http://www.ewg.org/about. Accessed October 2009.

_______. 2008f. Agriculture, in Past Accomplishments. http://www.ewg.org/about/accomplishments. Accessed October 2009.

_______. 2008g. What They Say about EWG's work. http://www.ewg.org/about/quotes. Accessed October 2009.

Ervin, David E., and Michael R. Dicks. 1988. Cropland Diversion for Conservation and Environmental Improvement: An Economic Welfare Analysis. *Land Economics* 64(3): 256–68.

Feather, Peter, and Daniel Hellerstein. 1997. Calibrating Benefit Function Transfer to Assess the Conservation Reserve Program. *American Journal of Agricultural Economics* 79(1): 151–62.

Feng, Hongli, Catherine L. Kling, and Philip W. Gassman. 2004. Carbon Sequestration, Co-benefits, ad Conservation Programs. Working Paper 04-WP 379. Ames IA, USA: Iowa State University, Center for Agricultural and Rural Development.

Feng, Hongli, Catherine L. Kling, Lyubov A. Kurkalova, and Silvia Secchi. 2003. Subsidies! The Other Incentive-Based Instrument: The Case of the Conservation Reserve Program. Working Paper 03-WP 345. Ames IA, USA: Iowa State University, Center for Agricultural and Rural Development.

Feng, Hongli, Catherine L. Kling, Lyubov A. Kurkalova, Silvia Secchi, and Philip W. Gassman. 2005. The Conservation Reserve Program in the Presence of a Working Land Alternative: Implications for Environmental Quality, Program Participation, and Income Transfer. Working Paper 05-WP 402. Ames IA, USA: Iowa State University, Center for Agricultural and Rural Development.

Food Agriculture Conservation and Trade Act of 1990, Part 4 of 11. 1990. Public Law 101-624, Nov. 28, 1990, 104 Stat. Available from http://www.nationalaglawcenter.org/assets/farmbills/1990-4.pdf. Accessed September 2009.

Gan, Jianbang, Okwudili O. Onianwa, John Schelhas, Gerald C. Wheelock, and Mark R. Dubois. 2005. Does Race Matter in Landowners' Participation in Conservation Initiative Programs? *Society and Natural Resources* 18: 431–45.

GAO (U.S. General Accounting Office) 1989. Farm Programs: Conservation Reserve Program Could Be Less Costly and More Effective. GAO/RCED-90-13 (November).

Effective July 7, 2004, the U.S. General Accounting Office changed its name to the U.S. Government Accountability Office.

_______. 1992. Conservation Reserve Program: Cost-Effectiveness Is Uncertain. GAO/RCED-93-132 (March).

_______. 1995a. Conservation Reserve Program: Alternatives Are Available for Managing Environmentally Sensitive Cropland. GAO/RCED-95-42 (February).

_______. 1995b. Farm Bill Issues. GAO/RCED-95-93R (February).

_______. 1998. Farm programs: Administrative requirements reduced and further program delivery changes possible. GAO/RCED-98-98 (April).

_______. 1999. Conservation Reserve Program Funding Requirements for the Natural Resources Conservation Service's Technical Assistance. AO/RCED-99-247R (August).

_______. 2002. Agricultural Conservation: State Advisory Committees' Views on How USDA Programs Could Better Address Environmental Concerns. GAO-02-295 (February).

_______. 2003. Climate Change: Preliminary Observations on the Administration's February 2002 Climate Initiative. GAO-04-131T (October).

_______. 2004. Farm Program Payments: USDA Needs to Strengthen Regulations and Oversight to Better Ensure Recipients Do Not Circumvent Payment Limitations. GAO-04-407 (April).

_______ (U.S. Government Accountability Office). 2005. Environmental Information: Status of Federal Data Programs That Support Ecological Indicators. GAO-05-376 (February).

Goodbody, Jerry. 2005. Green Acres. *Audubon Magazine* online, November 1. http://audubonmagazine.org/features0511/workingLands.html. Accessed September 2009.

Goodsell, Paul. 1995. Ag Processor Argues Against Set-Asides. *Omaha World Herald*, March 30, 1.

Gordon, Greg. 1995. Cropland Conservation Program Left in Limbo; Farm Bill negotiations could set its course. *Star Tribune* (Minneapolis, MN), October 30, 1A.

Hamilton, James T. 1995a. Pollution as News: Media and Stock Reactions to the Toxics Release Inventory Data. *Journal of Environmental Economics and Management*, 28: 98–113.

_______. 1995b. Testing for Environmental Racism: Prejudice, Profits, Political Power? *Journal of Policy Analysis and Management* 14 (1): 107–32.

_______. 1997. Taxes, Torts, and the Toxics Release Inventory: Congressional Voting on Instruments to Control Pollution. *Economic Inquiry* 35(4): 745–62.

_______. 1999. Exercising Property Rights to Pollute: Do Cancer Risks and Politics Affect Plant Emission Reductions? *Journal of Risk and Uncertainty* 18(2): 105–24.

_______. 2004. *All the News That's Fit to Sell: How the Market Transforms Information into News*. Princeton, NJ, USA: Princeton University Press.

_______. 2005. *Regulation through Revelation: The Origin, Politics, and Impacts of the Toxics Release Inventory Program*. New York: Cambridge University Press.

Hamilton, James T., and Scott de Marchi. 2006. Assessing the Accuracy of Self-Reported Data: An Evaluation of the Toxics Release Inventory. *Journal of Risk and Uncertainty* 13: 59–78.

Hamilton, James T., and Christopher H. Schroeder. 1994. Strategic Regulators and the Choice of Rulemaking Procedures: The Selection of Formal and Informal Rules Regulating Hazardous Waste. *Law and Contemporary Problems* 57: 111–60.

Hamilton, James T., and W. Kip Viscusi. 1999. *Calculating Risks? The Spatial and Political Dimensions of Hazardous Waste Policy*. Cambridge, MA, USA: MIT Press.

Hasten, Mike. 2007. Study: Farm runoff feeds dead zone. *Lafayette Daily Advertiser*, September 12.

Hendee, David, 1994. USDA Data on Erosion Questioned; Private group, AG officials are at odds. *Omaha World Herald*, November 23, 1.

Herszenhorn, David M. 2007. A Bid to Overhaul a Farm Bill Yields Subtle Changes. *New York Times*, 24 October.

Hughes, Jennie S., Dana L. Hoag, and Terry E. Nipp. 1995. The Conservation Reserve: A Survey of Research and Interest Groups. Special publication, *Council for Agricultural Science and Technology* 19 (July).

Ibendahl, Gregory. 2004. A Risk-Adjusted Comparison of Conservation Reserve Program Payments versus Production Payment for a Corn-Soybean Farmer. *Journal of Agricultural and Applied Economics* 36(2): 425–34.

Isik, Murat, and Wanhong Yang. 2004. An Analysis of the Effects of Uncertainty and Irreversibility on Farmer Participation in the Conservation Reserve Program. *Journal of Agricultural and Resource Economics* 29(2): 242–59.

Jesse's Hunting and Outdoors.com. 2008. Jesse's Hunting and Outdoor Forum. Conservation programs win with passage of farm bill; Congressional Sportsmen's Caucus Preserves Funding. May 16. Photocopy available from author. http://www.jesseshunting.com/ Accessed September 18, 2008.

Johnson, Barbara. 2005. Conservation Reserve Program: Status and Current Issues. CRS Report for Congress RS21613. Washington, DC: Library of Congress, Congressional Research Service.

Johnson, Douglas H., and Michael D. Schwartz. 1993. The Conservation Reserve Program and Grassland Birds. *Conservation Biology* 7(4): 934–7.

Kahneman, Daniel, Jack L. Knetsch, and Richard H. Thaler. 1991. Anomalies: The Endowment Effect, Loss Aversion, and Status Quo Bias. *Journal of Economic Perspectives* 5(1): 193–206.

Kiewiet, D. Roderick, and Mathew Daniel McCubbins. 1991. *The Logic of Delegation: Congressional Parties and the Appropriations Process*. Chicago: University of Chicago Press.

Kirby, Matt. 2008. Wild Blog: Conservation Reserve Program Safe for Now. August 1. Photocopy available from author. http://www.sierraclub.org/ Accessed 22 September 2008.

Klute, David S., Robert J. Robel, and Kenneth E. Kemp. 1997. Will Conversion of Conservation Reserve Program (CRP) Lands to Pasture Be Detrimental for Grassland Birds in Kansas? *American Midland Naturalist* 137(2): 206–12.

Lambrecht, Bill. 1994. Farmers Face Loss of Green Payments. *St. Louis Post-Dispatch*, April 8, 1A.

Lant, Christopher L., Steven E. Kraft, Jeffrey Beaulieu, David Bennett, Timothy Loftus, and John Nicklow. 2005. Using GIS-Based Ecological-Economic Modeling to Evaluate Policies Affecting Agricultural Watersheds. *Ecological Economics* 55: 467–84.

Lochhead, Carolyn. 2007. Yes, San Francisco Is the Land of Cotton Subsidies. *San Francisco Chronicle*, 23 September.

Lopez, Rigoberto A. 2001. Campaign Contributions and Agricultural Subsidies. *Economics and Politics* 13(3): 257–79.

Lubowski, Ruben N., Andrew J. Plantinga, and Robert N. Stavins. 2003. Determinants of Land-Use Change in the United States 1982–1997. RFF Discussion Paper 03-47. Washington, DC: Resources for the Future.

Madison, James. 1788. The Federalist No. 51. In *The Federalist Papers*. http://www.constitution.org/fed/federa51.htm. Accessed September 2009.

MAPLight.org. 2008. Remote Control: U.S. House Members raise 79% of campaign funds from outside their districts. http://www.maplight.org/remotecontrol08. Accessed September 2009.

Marshall, Bob. 1996. Outdoors Folks Deserve Credit for Farm Bill. *Times-Picayune*, April 7, C13.

McAllister, Bill. 1988. White House Forms Interagency Panel to Devise Plan to Cope with Drought. *Washington Post*, June 17, G1.

McCoy, Timothy D., Mark R. Ryan, and Loren W. Burger, Jr. 2001. Grassland Bird Conservation: CP1 vs. CP2 Plantings in Conservation Reserve Program Fields in Missouri. *American Midland Naturalist* 145(1): 1–17.

McCubbins, Mathew D., Roger G. Noll, and Barry R. Weingast. 1987. Administrative Procedures as Instruments of Political Control. *Journal of Law, Economics, and Organization* 3(2): 243–77.

McCubbins, Mathew D., and Thomas Schwartz. 1984. Congressional Oversight Overlooked: Police Patrols versus Fire Alarms. *American Journal of Political Science* 28(1): 165–79.

Miranda, Marie Lynn. 1992. Landowner Incorporation of Onsite Soil Erosion Costs: An Application to the Conservation Reserve Program. *American Journal of Agricultural Economics* 74(2): 434–43.

Mitchell, Larry. 2008. Herger Thinks Cutting Conservation Funds in New Farm Bill Is All Right. *Chico Enterprise-Record*, April 4.

New York Times. 2000. The 2000 Campaign; Exchanges between the candidates in the third Presidential debate. October 18, A26.

Nomsen, D., and Mark Herwig. 2005. Conservation Reserve Program: Two decades of success. Photocopy available from author. http://www.quailforever.org/conservation/?full=y. Accessed February 3, 2006.

O'Donnell, Kim. 2007. So What's This Farm Bill? A Mighty Appetite, *Washington Post Blogs*, July18.

Olson, Allen H. 2001. Lead commentary: Federal Farm Programs—Past, Present, and Future. Will we learn from our mistakes? *Great Plains Natural Resources Journal* 6 (Fall 2001), 1–29.

OpenSecrets.org. Big Picture: Price of Admission. http://www.opensecrets.org/bigpicture/stats.php?cycle=2006&Type=R&Display=A. Accessed September 2009.

Osborn, C. Tim, Felix Llacuna, and Michael Linsenbigler. 1995. The Conservation Reserve Program: Enrollment statistics for Signup Periods 1–12 and Fiscal Years 1986–93. *USDA Statistical Bulletin*, no. SB925 (November).

Osborn, T. 1993. The Conservation Reserve Program: Status, Future, and Policy Options. *Journal of Soil and Water Conservation* 48(4): 271–8.

Parkhurst, Gregory M., and Jason F. Shogren. 2003. Evaluating Incentive Mechanisms for Conserving Habitat. *Natural Resources Journal* 43: 1093–1150.

Parks, Peter J., and Ian W. Hardie. 1995. Least-Cost Forest Carbon Reserves: Cost-Effective Subsidies to Convert Marginal Agricultural Land of Forests. *Land Economics* 71(1): 122–36.

Parks, Peter J., and James P. Schorr. 1997. Sustaining Open Space Benefits in the Northeast: An Evaluation of the Conservation Reserve Program. *Journal of Environmental Economics and Management* 32: 85–94.

Patterson, Matthew P., and L. B. Best. 1996. Bird Abundance and Nesting Success in Iowa CRP Fields: The Importance of Vegetation Structure and Composition. *American Midland Naturalist* 135(1): 153–67.

Perez, Michelle. 2007. Trouble Downstream: Upgrading Conservation Compliance. Environmental Working Group Research. http://www.ewg.org/node/22513. Accessed September 2009.

Porter, Larry. 1995. Pheasant Hunter Likes January Best. *Omaha World Herald*, January 29, 11C.

_______. 1997. Environmental Benefits Index Now Rates CRP Land. *Omaha World Herald*, October 26, 6C.

_______. 2003. Johanns Targets Pheasants. *Omaha World Herald*, February 16, 10C.

Powell, Mark R., and James D. Wilson. 1997. Risk Assessment for National Natural Resource Conservation Programs. RFF Discussion Paper 97–49. Washington, DC: Resources for the Future.

ProgressiveConservative.com. 2008. From the vault: Conservation Reserve Program in Trouble, August 21. Posted on "The Big Stick" blog. http://thebigstick.wordpress.com/2008/08/21/conservation-reserve-program-in-trouble/

Reichelderfer, Katherine, and William G. Boggess. 1988. Government Decision Making and Program Performance: The Case of the Conservation Reserve Program. *American Journal of Agricultural Economics* 70(1): 1–11.

Reuters. 2007. U.S. Farmers Should Curb Fertilizer Runoff: Study. September 10. http://www.reuters.com/article/environmentNews/idUSN1033310520070910. Accessed September 2009.

_______. 2007. Ag Secretary Urged to Reject Early Release of Land in Conservation Reserve Program. July 9.

Richert, Catharine. 2008. Farm Bill Highlights. *Congressional Quarterly Today*, May 8.

Ribaudo, Marc O. 1989. Targeting the Conservation Reserve Program to Maximize Water Quality Benefits. *Land Economics* 65(4): 320–32.

_______. 2004. Consideration of Offsite Impacts in Targeting Soil Conservation Programs. In *Economics of Agri-environmental Policy*. Vol. 2. Edited by Sandra S. Batie and Richard D. Horan. Burlington, VT, USA: Ashgate.

Ribaudo, Marc O., Dana L. Hoag, Mark E. Smith, and Ralph Heimlich. 2001. Environmental Indices and the Politics of the Conservation Reserve Program. *Ecological Indicators* 1: 11–20.

Ribaudo, Marc O., Steven Piper, Glenn D. Schaible, Linda L. Langer, and Daniel Colacicco. 2004. CRP: What Economic Benefits? In *Economics of Agri-environmental Policy*. Vol. 2. Edited by Sandra S. Batie and Richard D. Horan. Burlington, VT, USA: Ashgate.

Robbins, William. 1987. Anti-erosion Program for Farms Accelerates. *New York Times*, March 14, section 1, 7.

Roberts, Michael J., and Shawn Bucholtz. 2005. Slippage in the Conservation Reserve Program or Spurious Correlation? A Comment. *American Journal of Agricultural Economics* 87(1): 244–50.

Robles, Marcos D., and Ingrid C. Burke. 1997. Legume, Grass, and Conservation Reserve Program Effects on Soil Organic Matter Recovery. *Ecological Applications* 7(2): 345–57.

Rocky Mountain News. 2002. 2002 Farm Bill Increases Funding for Conservation. May 31, 17C.

St. Louis Post-Dispatch. 1989a. Farmers Paid for Land Set Aside to Combat Erosion (part 1 of a 5-part series). May 9, 6.

_______. 1989b. Bill Provides Owners 'Motivation' for Saving Land. *St. Louis Post Dispatch*, June 6, 7.

Schara, Ron, and Dennis Anderson. 1994. Congress Is Urged to Spare the CRP; Plan cited as benefiting wildlife, improving soil, lifting grain prices. *Star Tribune* (Minneapolis, MN), April 27, 9C.

Shaw, Hank. 2008. Farm Bill Fight Intensifies: Conservation funds now focus of finalizing deal. *Record* (Stockton, CA), April 4.

Shoemaker, Robbin. 1989. Agricultural Land Values and Rents under the Conservation Reserve Program. *Land Economics* 65(2): 131–7.

Siegel, Paul B., and Thomas G. Johnson. 1991. Break-Even Analysis of the Conservation Reserve Program: The Virginia Case. *Land Economics* 67(4): 447–61.

Smith, Doug. 1997. Farmers Flock to CRP, but Not All Will Qualify. *Star Tribune* (Minneapolis, MN), March 16, 20C.

_______. 2004. Working for Wildlife; Group's voice is heard nationally. *Star Tribune* (Minneapolis, MN), May 3, 11A.

Smith, Rodney B.W. 1995. The Conservation Reserve Program as a Least-Cost Land Retirement Mechanism. *American Journal of Agricultural Economics* 77(1): 93–105.

Sternstein, Aliya. 2008. Livestock Grazing On Conserved Land Welcomed by Farm-State Lawmakers. *Congressional Quarterly Today*, May 28.

Stratmann, Thomas. 1992a. Are Contributors Rational? Untangling Strategies of Political Action Committees. *Journal of Political Economy* 100(3): 647–64.

______. 1992b. The Effects of Logrolling on Congressional Voting. *American Economic Review* 82(5): 1162–76.

_______. 1995. Campaign Contributions And Congressional Voting: Does the Timing of Contributions Matter? *Review of Economics and Statistics* 77(1): 127–36.

_______. 1998. The Market for Congressional Votes: Is Timing of Contributions Everything? *Journal of Law and Economics* 41(1): 85–113.

Streitfeld, David. 2008. As Prices Rise, Farmers Spurn Conservation Program. *New York Times*, April 9.

Swanson, D. A., D. P. Scott, and D. L. Risley. 1999. Wildlife Benefits of the Conservation Reserve Program in Ohio. *Journal of Soil and Water Conservation* 54(1): 390–4.

USDA (U. S. Department of Agriculture). Agricultural Stabilization and Conservation Service, Commodity Credit Corporation. 1986. Conservation Reserve Program. 7 CFR Part 704, interim rule. Washington, D.C.

_______. Agricultural Stabilization and Conservation Service, Commodity Credit Corporation. 1987. Conservation Reserve Program. 52 FR, Part 4265, final rule. Washington, DC.

_______. Commodity Credit Corporation. 2003. 2002 Farm Bill Conservation Reserve Program. Long Term Policy. 7 CFR, Part 1410, part IV, 68 FR 24830, interim rule. Washington, DC.

_______. Commodity Credit Corporation. 2004. 2002 Farm Bill Conservation Reserve Program. Long Term Policy. 7 CFR, Part 1410, 69 FR 26755, final rule. Washington, DC.

_______. Economic Research Service. 1997. New CRP Criteria Enhance Environmental Gains. *Agricultural Outlook* (October).

_______. Farm Service Agency. 1997. Conservation Reserve Program: Long-Term Policy. 7 CFR, Parts 704 and 1410, part III, 62 FR7602, final rule. Washington, DC.

_______. Farm Service Agency. 2007. Conservation Reserve Program: Summary and Enrollment Statistics FY 2006. Washington, DC.

_______. Farm Service Agency. 2008. Conservation Reserve Program: Monthly Summary— October 2008. Washington, DC. http://www.fsa.usda.gov/Internet/FSA_File/oct2008.pdf. Accessed September 2009.

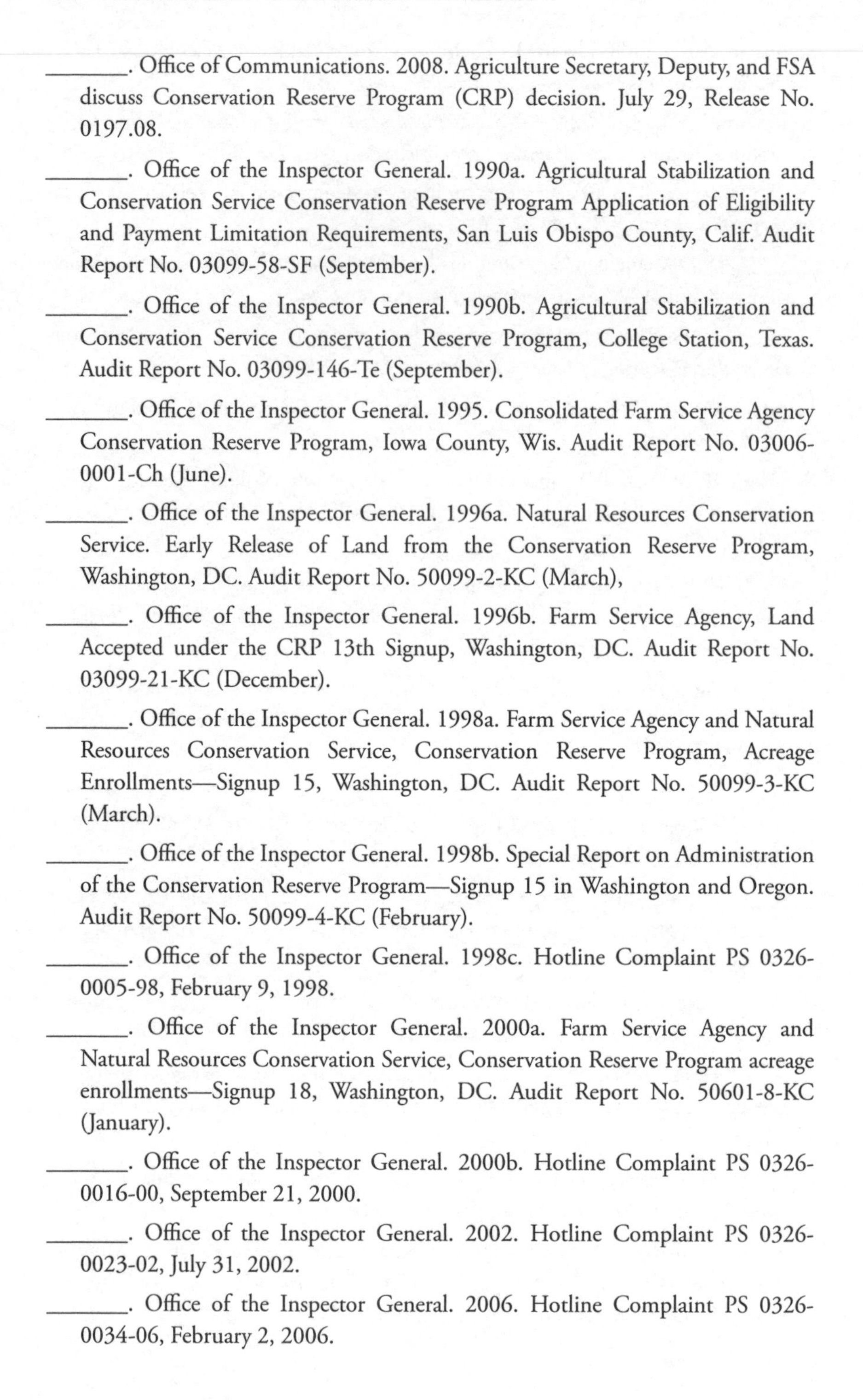

_______. Office of Communications. 2008. Agriculture Secretary, Deputy, and FSA discuss Conservation Reserve Program (CRP) decision. July 29, Release No. 0197.08.

_______. Office of the Inspector General. 1990a. Agricultural Stabilization and Conservation Service Conservation Reserve Program Application of Eligibility and Payment Limitation Requirements, San Luis Obispo County, Calif. Audit Report No. 03099-58-SF (September).

_______. Office of the Inspector General. 1990b. Agricultural Stabilization and Conservation Service Conservation Reserve Program, College Station, Texas. Audit Report No. 03099-146-Te (September).

_______. Office of the Inspector General. 1995. Consolidated Farm Service Agency Conservation Reserve Program, Iowa County, Wis. Audit Report No. 03006-0001-Ch (June).

_______. Office of the Inspector General. 1996a. Natural Resources Conservation Service. Early Release of Land from the Conservation Reserve Program, Washington, DC. Audit Report No. 50099-2-KC (March),

_______. Office of the Inspector General. 1996b. Farm Service Agency, Land Accepted under the CRP 13th Signup, Washington, DC. Audit Report No. 03099-21-KC (December).

_______. Office of the Inspector General. 1998a. Farm Service Agency and Natural Resources Conservation Service, Conservation Reserve Program, Acreage Enrollments—Signup 15, Washington, DC. Audit Report No. 50099-3-KC (March).

_______. Office of the Inspector General. 1998b. Special Report on Administration of the Conservation Reserve Program—Signup 15 in Washington and Oregon. Audit Report No. 50099-4-KC (February).

_______. Office of the Inspector General. 1998c. Hotline Complaint PS 0326-0005-98, February 9, 1998.

_______. Office of the Inspector General. 2000a. Farm Service Agency and Natural Resources Conservation Service, Conservation Reserve Program acreage enrollments—Signup 18, Washington, DC. Audit Report No. 50601-8-KC (January).

_______. Office of the Inspector General. 2000b. Hotline Complaint PS 0326-0016-00, September 21, 2000.

_______. Office of the Inspector General. 2002. Hotline Complaint PS 0326-0023-02, July 31, 2002.

_______. Office of the Inspector General. 2006. Hotline Complaint PS 0326-0034-06, February 2, 2006.

_______. Office of the Inspector General. 2007. Hotline Complaint PS 0326-0036-07, February 9, 2007.

_______. Office of the Inspector General. 2008. Hotline Complaint PS 0326-0039-07, April 19, 2007.

U.S. Department of the Interior. U. S. Geological Survey. 2003. A National Survey of Conservation Reserve Program (CRP) Participants on Environmental Effects, Wildlife Issues, and Vegetation Management on Program Lands. *Biological Sciences Report*, USGS/BRD/BSR-2003-0001.

U.S. House of Representatives. 1994. Subcommittee on Environment, Credit, and Rural Development of the Committee on Agriculture. Review of the Budget and Policy Consequences of Extending the Conservation Reserve Program. Hearing before the Subcommittee on Environment, Credit, and Rural Development of the Committee on Agriculture. 103rd Cong., second session, Aug. 2, 1994.

———. 1996. Committee on Agriculture. Hearing before the Committee on Agriculture. 104th Cong., Sep. 19, 1996.

———. 1997a. Subcommittee on Forestry, Resource Conservation and Research of the Committee on Agriculture. Hearing before the Subcommittee on Forestry, Resource Conservation and Research of the Committee on Agriculture. 105th Cong., first session, Feb. 26, 1997.

———. 1997b. Subcommittee on Forestry, Resource Conservation and Research of the Committee on Agriculture. Hearing before the Subcommittee on Forestry, Resource Conservation and Research of the Committee on Agriculture. 105th Cong., first session, June 11, 1997.

———. 1999. Subcommittee on General Farm Commodities, Resource Conservation, and Credit of the Committee on Agriculture. Hearing before the Subcommittee on General Farm Commodities, Resource Conservation, and Credit of the Committee on Agriculture. 106th Cong., first session, Jul 22, 1999.

———. 2000. Subcommittee on General Farm Commodities, Resource Conservation, and Credit of the Committee on Agriculture. Hearing before the Subcommittee on General Farm Commodities, Resource Conservation, and Credit of the Committee on Agriculture. 106th Cong., second session, Mar. 31, 2000.

U.S. House of Representatives and U.S. Senate. 1994. Subcommittee on Environment, Credit, and Rural Development of the Committee on Agriculture, U.S. House of Representatives and the Subcommittee on Agricultural Research, Conservation, Forestry, and General Legislation of the Committee on Agriculture, Nutrition, and Forestry, U.S. Senate. Joint hearing before Subcommittee on Environment, Credit, and Rural Development of the Committee on Agriculture, U.S. House of

Representatives and the Subcommittee on Agricultural Research, Conservation, Forestry, and General Legislation of the Committee on Agriculture , Nutrition, and Forestry, U.S. Senate. 103th Cong., second session, September 1, 1994.

U.S. Newswire. 1997. Environmental Working Group Statement on Conservation Reserve Program. May 27, 1997.

U.S. Senate. 1988. Subcommittee on Conservation and Forestry of the Committee on Agriculture, Nutrition, and Forestry. Hearing before the Subcommittee on Conservation and Forestry of the Committee on Agriculture, Nutrition, and Forestry. 100th Cong., second session, January 19, 1988.

U.S. Senate. 2006. Committee on Agriculture, Nutrition, and Forestry. Hearing before the Committee on Agriculture, Nutrition, and Forestry. 109th Cong., second session, June 7, 2006.

Van Doren, Terry D., Dana L. Hoag, and Thomas G. Field. 1999. Political and Economic Factors Affecting Agricultural PAC Contribution Strategies. *American Journal of Agricultural Economics* 81(2): 397–407.

Viscusi, W. Kip, and James T. Hamilton. 1999. Are Risk Regulators Rational? Evidence From Hazardous Waste Cleanup Decisions. *American Economic Review* 89(4): 1010–27.

White House. 2004. Conservation Initiatives Fact Sheet. 4 August. Available from http://georgewbush-whitehouse.archives.gov/news/releases/2004/08/20040804-14.html

Wu, Jun Jie. 2000. Slippage Effects of the Conservation Reserve Program. *American Journal of Agricultural Economics* 82(4): 979–92.

______. 2005. Slippage Effects of the Conservation Reserve Program. A Reply. *American Journal of Agricultural Economics* 87(1): 251–4.

Wuerthner, George. 2008. *Redstaterebels.com* website. Conservation Reserve Program: Government Boondoggle. Posted July 24, 2009. http://redstaterebels.org/2008/07/boondoggle-in-the-fields. Accessed September 2009.

Yang, Wanhong, Madhu Khanna, Richard Farnsworth, and Haryi Onal. 2005. Is Geographical Targeting Cost-Effective? The Case of the Conservation Reserve Enhancement Program in Illinois. *Review of Agricultural Economics* 27(1): 70–88.

Young, C. Edwin, and C. Tim Osborn. 1990. Costs and Benefits of the Conservation Reserve Program. *Journal of Soil and Water Conservation* 45(3): 370–3.

Yu, Zhihao. 2005. Environment Protection: A Theory of Direct And Indirect Competition for Political Influence. *Review of Economic Studies* 72(1): 269–86.

Zinn, Jeffrey. 2003. Conservation Reserve Program: Status and Current Issues. *National Library for the Environment* (May 23).

INDEX